Communicating Science in Public

Megan Shields Formato, Kevin C. Moore,
Sarah Peterson Pittock, Emily Polk,
and Cassie A. Wright

CRC Press
Taylor & Francis Group
Boca Raton London New York

CRC Press is an imprint of the
Taylor & Francis Group, an **informa** business

Designed cover image: Shutterstock

First edition published 2027
by CRC Press
2385 NW Executive Center Drive, Suite 320, Boca Raton FL 33431

and by CRC Press
4 Park Square, Milton Park, Abingdon, Oxon, OX14 4RN

CRC Press is an imprint of Taylor & Francis Group, LLC

ISBN: 9781032860114 (hbk)
ISBN: 9781032848648 (pbk)
ISBN: 9781003520887 (ebk)

DOI: 10.1201/9781003520887

Typeset in Minion
by codeMantra

Contents

Acknowledgments

THE COLLABORATION FROM WHICH this book emerged extends well beyond the list of co-authors. We offer our gratitude to all the science communicators whose frameworks, theories, and practices we platform in this book and who have inspired our own ideas for how to advance the work. The book benefited immensely from the collaborative spirit, scholarship, and teaching in Stanford's Program in Writing and Rhetoric, and, especially, from our colleagues who teach with us within the Notation in Science Communication program.

We wish to particularly acknowledge Dr. Shay Brawn and Dr. Jennifer Stonaker, who first developed the Introduction to Science Communication course that inspired and informed the structure and approach of this book. Dr. Clara Lewis first developed the course Self and Science that inspired our chapter on using personal narratives to communicate science. We would also like to thank our science communication students, whose interest in working in new genres pushed us to develop new materials and courses and to think carefully about what it means to communicate science in public today. We particularly acknowledge the students in the Winter 2026 Introduction to Science Communication and Data Visualization courses for being some of our first readers and offering excellent feedback as we revised the book manuscript.

We were grateful to receive a research grant from the Program in Writing and Rhetoric, which supported early work on the book and made it possible for us to travel to the Conference on College Composition and Communication in 2025 and present work in progress to colleagues who share our commitment to teaching science communication at other institutions. We also wish to thank the editorial team at Taylor & Francis for their guidance and support and the anonymous reviewers of our book proposal who offered essential and generous feedback.

About the Authors

Megan Shields Formato, PhD, is a historian of modern science and technology whose research focuses on the history of science communication and the role of women in science. She is an Advanced Lecturer in Stanford University's Program in Writing and Rhetoric, where she teaches courses in science communication and on the relationship between technology and culture. Her writing has appeared in *Science, American Scientist, Endeavor, Lady Science,* and the *AAEEBL ePortfolio Review* and in edited collections including *Embodied Knowledge: Women in Science & Technology before Silicon Valley* and *Beyond Fitting In: Rethinking First-Generation Writing and Literacy Education.*

Kevin C. Moore, PhD, is an Advanced Lecturer in the Program in Writing and Rhetoric (PWR) at Stanford University, where he teaches writing and science communication. Previously, he taught writing courses for engineers and scientists at the University of California, Santa Barbara. Trained in American literature and rhetoric, his current research interests lie at the intersections of science communication and propaganda studies. His work has appeared in *Arizona Quarterly, Arts, ContraSTS, Writing on the Edge, African American Review, Composition Studies, MAKE,* and *Souciant*, as well as collections such as *Trigger Warnings: Teaching through Trauma* and *Creative Ways of Knowing in Engineering.*

Sarah Peterson Pittock, PhD, is the Associate Director of Civic, Liberal, and Global Education (COLLEGE) at Stanford University. Previously, she served as an Advanced Lecturer in the Program in Writing and Rhetoric at Stanford University, where she taught courses in research writing, public speaking, tutor education, and science communication. Her work has been published in *Women's Writing, The WAC Journal, Writing Center Journal, Writing on the Edge*, and edited collections.

Emily Polk, PhD, teaches and writes about community-led responses to climate change, the mobilization of social movements, and inclusive science communication at Stanford University. She developed and taught some of the first courses at Stanford on Communicating Climate Change and Environmental Justice Storytelling. Prior to getting her doctorate, she worked as a human rights- and environment-focused writer and editor for nearly ten years, helping to produce radio documentaries in Burmese refugee camps and facilitating a human-rights-based newspaper in a Liberian refugee camp. Her first book is *Communicating Global to Local Resiliency: A Case Study of the Transition Movement.*

Cassie A. Wright, PhD, is an Advanced Lecturer in the Program in Writing and Rhetoric (PWR) at Stanford University, where she teaches courses in science communication. She also teaches courses on censorship and social media and their impacts on modern literacy and diversity of thought. Her work has appeared in *Rhetoric Review* and the *Journal of Writing Program Administration*, and she is the author of the writing studies textbook *Sport* and a contributing author to the writing teacher's guide *Assigning, Responding, Evaluating,* 5th edition.

PART 1

Why Communicate Science in Public Now?

CHAPTER 1

Introduction

WHY COMMUNICATING SCIENCE *IN PUBLIC*

We are always potentially communicating science in public. Yes, there are institutional perimeters within science, for instance, journal paywalls and the gatekeeping structures of peer review. Technical jargon, as well as the sheer rigor of specialist discourse—that is to say, communication among scientists—can also contribute to the impression that scientific research spaces are a closed, siloed world. But if science ever appears to be a black-box domain for discovering and brokering truths that will later be revealed to general audiences, this is an illusion. The reality is that public-facing science communication always exists on a continuum with research. Even proprietary or secretive research by companies and governments affects the public and is the public's business, although the timeline by which that work is made public may be longer or shaped by forces trying to control a narrative or hype and thus sell a finding or product. And even though good science communication indeed needs to pursue strategies for navigating technical opacity and public resistance to unwelcome truths, because science is inherently public, science must pursue the public good—or else what is it for?

Effective science communication involves the work of translating opaque or complex concepts for interested non-specialist audiences, but it neither begins nor ends there. Rather, science communication is a process of participating in democratic deliberation and broadening our collective knowledge, with the goals of improving human lives and planetary health, as well as including those audiences and groups historically excluded from the conversation. At the outset of this book, we would like to invite you

DOI: 10.1201/9781003520887-2

to consider our title, *Communicating Science in Public*, to be not a special situation for scientists and others seeking to communicate science to non-expert audiences, but a redundant reminder that science is always already the domain of a diverse public.

Of course, a specific, global, sustained event in recent history made the public nature of science communication readily apparent to most people: the explosion of interest in science-informed policy that emerged in 2020 during the COVID-19 pandemic. Amid the pandemic, "trust the science" became at once a rallying slogan for solidarity and a sometimes-abused cliché. While the primary topics driving this interest were the questions of how and where to disseminate emerging findings regarding the pandemic and how to use those findings to mitigate it, the downstream effects of the COVID-19 moment on the genres and media forms by which we engage with science have touched all fields, and they persist. As one pair of scientists predicted at the outset of COVID-19, the pandemic was

> a supernova in human history: an expanding, all-encompassing set of events and responses to them that touch[ed] every aspect of the human condition...Over the next few decades, economists, epidemiologists, public health experts, historians, philosophers, sociologists, physicians, psychologists, and others will work to untangle the interwoven threads.
>
> *(Prasad & Flier, 2020)*

We would add science communicators to this list of public actors with a special role in this work of "untangling." The COVID-19 moment remade the science communication landscape. But the technologies, practices, and pressures that have shaped this new landscape have been in development for decades. If the pandemic was a historical "supernova," for science communicators it also became an amplifying lens through which publication and media trends, as well as broader epistemological challenges and crises—that is to say, challenges and crises in our understanding of how we know what we know—have been refracted into new forms. If, before the pandemic, many fields were grappling with a replication crisis engendered by institutional factors such as publication bias (i.e., the pressure scientists feel to publish for tenure, funding, and other forms of recognition), post-pandemic, dubious AI-written "slop" science has eroded not only internet sources and journalism but also the imperfect peer-review process itself (Andersen, 2026). If, in the mid- to late 2010s, fears about rampant, brazen data theft by tech companies and the rise of what Shoshana Zuboff calls "surveillance capitalism" constituted some of the most urgent risks presented by the digital age (where "surveillance capitalism" is defined as "[a] new economic order that claims human experience as free raw material for hidden commercial practices of extraction, prediction, and sales" as well as a "movement that aims to impose a new collective order based on total certainty" (Zuboff, 2019)), post-pandemic, the AI industry bears the fruit of precisely the robust harvest of data Zuboff and others warned about, dubiously remodeling, remixing, and reselling our collective experience and knowledge to us as an inescapable new channel of truth.

Simultaneously, the political fractures that occurred in the US and around the world in tandem with, and in some cases because of, the COVID-19 pandemic have been dramatic, persistent, and profoundly impactful to science. If, in 2016, you wanted to convince vaccine-hesitant parents to inoculate their children, you would likely have been reaching out to specific and usually very small religious communities, or perhaps a few isolated naturopathic or libertarian enclaves. In 2026, vast swaths of people in the United States oppose vaccines as a political matter, to the point where localized measles outbreaks have reemerged for the first time in decades; according to a widely cited Pew poll, in the US, the percentage of people who say they have little to no trust in scientists has hovered around 25% in the years following the pandemic, more than ten points higher than it was pre-pandemic (Tyson & Kennedy, 2024). Beginning in 2025, sweeping attacks on public funding for science at agencies such as the National Institutes of Health and the National Science Foundation can partly be considered a backlash to perceived overreach during the pandemic, although they have also transpired alongside decades-long trends toward the privatization of science in fields ranging from biomedical research to space exploration.

These are but a few examples taken from the realms of epistemology, technology, and politics, but they indicate how the challenges facing effective science communication are significant and rapidly evolving. Yet one assumption this volume takes for granted is that effective science communication is always hard—an affair of untangling signal from noise—and built on earned (and in some cases re-earned) trust. A corollary assumption is that successfully navigating the challenges of science communication requires learning to identify the features of rhetorical situation. We'll say more about rhetorical situation—as cogently outlined by the rhetorician Lloyd Bitzer in 1968—in the next chapter, but for the moment, you should think about it as an act of communication with a particular audience that is constrained by time, place, and genre. Effective science communication depends upon paying attention to these factors, as well as to the conventions, limitations, and opportunities offered by specific science communication genres, which can shift in value and significance in relation to their position in a broad ecology of ideas, genres, and media. We borrow the term ecology here from the idea in media studies of the "media ecology," where, when a new communications medium emerges in response to technological advances, other media in the ecology or landscape adapt and respond (Postman, 2000). Think, for instance, of how the arrival of podcasts as a medium changes the reasons why you might turn to long-form magazine articles or radio broadcasts: those related media still exist and matter, but they come to occupy a different place and purpose in the broader ecology. A podcast might be where you go to hear a dynamic, preliminary discussion of an issue; for a more rigorous, careful assessment, you might instead appreciate a long-form magazine piece (for one thing, it can be harder to take notes from a podcast, which is frequently consumed while multitasking at the gym, commuting, or doing chores).

Most of this book will be spent providing rhetorical tools, as well as genre-specific guidance, to help science communicators respond to the exigence culminated and precipitated by the COVID-19 moment and the "infodemic" it produced. Because identifying and naming exigence is a core principle of rhetorical thinking, however, we'd like to begin by mapping specifically how COVID-19 reshaped science communication in public and what it means to take up this work now. Indeed, one can hardly imagine a more exigent moment than the pandemic and its aftermath for science communication (where exigence was, for Bitzer, a feature of "almost any sort of context": "an imperfection marked by urgency; it is a defect, an obstacle, something waiting to be done, a thing which is other than it should be" (1968, p. 6)).

THE RISE OF PREPRINTS, THE DEATH OF TWITTER, THE MIRAGE OF AI: AN INSTRUCTIVE EPISODE IN SCIENCE COMMUNICATION AND A PERSISTENT "INFODEMIC"

In February 2020, as the world was beginning to respond to the outbreak of the COVID-19 pandemic with dramatic social restrictions, border closures, lockdowns, and other public health measures—all actions taken with the ostensible goal of following emerging science on the virus—the risk communication team of the World Health Organization launched a new communications platform called the WHO Information Network for Epidemics. This platform was meant to combat what WHO Director-General Tedros Adhanom Ghebreyesus had called, a few days earlier, a potential parallel "infodemic" of misinformation and, just as importantly, help the public access the most up-to-date, credible, science-based information to protect their health (2020). According to Sylvie Briand, Director of Infectious Hazards Management at WHO's Health Emergencies Programme, the goal of the platform was to address the threat of an infodemic running parallel to the actual epidemic was, on the one hand, nothing new: it has long been known (since the Middle Ages, according to Briand) that a "tsunami of information" accompanies significant outbreaks of disease, including both credible and dubious forms of information (Zarocostas, 2020). Yet for Briand, the COVID-19 moment was categorically different from previous epidemics because of the arrival of new technologies and media. "With social media" in particular, she told *The Lancet* in late February 2020,

> this phenomenon is amplified, faster and further, like the viruses that travel with people ... So it is a new challenge, and the challenge is the [timing] because you need to be faster if you want to fill the void.
>
> *(Zarocostas, 2020)*

For anyone interested in science communication who lived through the subsequent months and years of the extended COVID-19 crisis, the early efforts of the WHO and other health authorities and organizations to shape the public conversation on the pandemic around credible scientific information—to communicate swiftly to "fill the void"—will probably seem an entirely reasonable, well-intentioned, but unfortunately ill-fated gesture. Not only did malicious misinformation and propaganda about the virus run rampant in social media contexts, despite efforts to guard the public against it, but changing guidance responding to legitimate new findings about the disease also left the public overwhelmed with often contradictory guidance and knowledge. Metaphors, hypotheticals, and figures meant to deliberate health policy were often circulated wildly out of context on social media, where they could take on devious lives of their own. Simultaneously, with many people socially distanced and working remotely—including most members of the world's professional class responsible for maintaining educational and research institutions—new media, genres, and other forms of science communication proliferated not only to "fill the void," but in many cases to take advantage of the profound exigence the pandemic created to say something about the situation. (One retrospective meta-analysis found that, between 2020 and 2022, 3.7% of all published English-language research was on the topic of COVID-19 (Whitaker et al., 2025).) In the process, new, often hybrid media forms and pathways emerged, where the proximity of research, policy debate, journalism, and education played out in unprecedented configurations and conversations. More people were paying attention to science than ever before. Physicians and public health experts became celebrities overnight as their social media accounts became opportunities to make "facts" about the virus "go viral," as one Wall Street Journal headline punned early in the crisis ("Doctors Are Tweeting About Coronavirus to Make Facts Go Viral," Wells 2020); other scientists launched podcasts or penned op-eds for the first time during the pandemic. Yet this took place against a background in which the evolving science on the virus was also far more variable in character and quality, to the point where scientists themselves—not to mention the public—could not always keep up with the changing information landscape (let alone who and what forces it was ultimately serving).

One instructive site where a profound remixing of media and genres took place was in the radical expansion of, and attention given to, preprint servers, where preliminary research that previously might have been conducted relatively outside of the public's eye suddenly became visible and part of the supercharged public conversation. For students of science communication, the explosion of preprints in 2020 might be seen as the infodemic in microcosm. Whereas fields like physics and computer science had been using the preprint server arXiv since 1991 to share original research in order to provide authors with a means to trademark ideas, as well as to receive feedback in advance of submitting for peer review (and outside institutional paywalls), servers for other fields only appeared more recently. In the biomedical sciences, preprint servers were launched decades later—bioRxiv in 2013 and medRxiv in 2018—but their use did not become mainstream until 2020, amid the urgency to discover, publish, and disseminate important findings about the emergent virus. Suddenly, in 2020, journalists, politicians, technology moguls, armchair epidemiologists, as well as researchers themselves, were engaging with a scientific interface where traditional institutional guardrails were abruptly reshaped. As Twitter feeds were integrated directly into the bioRxiv and medRxiv platforms in 2020, participants in the conversation—from senior research faculty to graduate students new to their fields, from the average person sheltering in place at home to celebrities with enormous followings—could share preliminary findings published in advance of peer review directly to social media, which became a primary site of scientific debate. A little like the arrival of ticker-tape newsreels at the bottom of the screen during cable and broadcast television news coverage in the wake of the 9/11 attacks, which proliferated to report the surfeit of information attending that emergency, preprint servers, as well as practices for reporting on and responding to them, have remained and become lasting elements of the media ecology despite the passing of the initial crisis.

Just as the COVID-19 virus remains with us as an endemic affliction, and just as preprints are now a permanent part of the media and scientific publishing landscape, the broader urgency and instability, as well as the very public nature, of the "infodemic" persist. Again, COVID-19's impact on science communication was the culmination of forces that had been coalescing for years, but the infodemic has become the normal background state science communicators must confront. Preprint servers may have since downgraded the visibility of their direct integration of social media feeds into their platforms, but they have become lasting pathways for disseminating research and a place where science journalists and other science communicators look for stories about emerging findings. After Twitter was purchased amid considerable controversy by Elon Musk in 2022 and, in 2023, rebranded as X, the online communities of "Scientific Twitter" largely collapsed, but they have now migrated elsewhere, to federated forums such as BlueSky or Mastodon (this migration extends far beyond the biomedical sciences, to all academic disciplines). Our situation, as communicators trying to parse an ever-shifting media and genre landscape, becomes an affair of navigating what the media theorist Anna Kornbluh (2024) has called a state of "immediacy," where we no longer know—and perhaps cannot know—where one medium or platform begins and ends.

After COVID-19, we expect the ground to shift beneath our feet. Of course, the situation is complicated further by the arrival of extractive "generative AI" technologies, which undermine copyright and contribute to misinformation. Although mainstream attention to AI didn't emerge until the fall of 2022 with the rollout of OpenAI's ChatGPT tool, generative

AI can better be thought of not as what followed the infodemic, but as part of it. AI is an extension of the infodemic. While AI may have certain discrete uses within scientific research, for those who care about science communication, the dangers of AI are many. Relying on it to assist with content blunts the development of your rhetorical skills, circumventing integral steps like developing an awareness of your audience's interests, knowledge, and beliefs. AI might also lead to misleading or incomplete information when used as a research tool. More broadly, because AI is designed to present information in the tone of authoritative truth, using these systems disrupts the credibility of all science communication by chipping away at what philosopher Regina Rini has called, in her research on the related problem of deceptive deepfake video technology, the "epistemic backstop" (2020). The epistemic backstop is our ability to rely consistently on established media and practices related to evidence in order to make arguments and ultimately come to a consensus about what we know. As science historian Alex Csiszar notes, genres of modern science communication are particularly tightly bound up with attribution conventions, offering ways of clearly indicating to readers where ideas and evidence come from, and who should take credit and responsibility. But

> [t]ext generated by ChatGPT might be indebted both to a vast array of previous authors and to no one in particular at the same moment, with no easy way to discern the details. This is obliteration by incorporation at a whole other level.
>
> *(2024, p. 615)*

Science communication depends fundamentally on evidence and attribution, and just as fundamentally on the assumption that humans are the ones communicating the science, not mathematical models generating statistically likely approximations of science from a corpus of texts that might be authored by anyone or no one, or that could even draw from retracted scientific articles (Ananya, 2025).

While the long-term impacts of AI on the media ecologies of science communication remain unsettled, against the enormous hype promoted by the profit-driven AI industry in the years since its initial rollout, science communicators have a responsibility to recognize AI's role as fundamentally corrosive to trust. Throughout this book, we offer you ways to create and sustain trust with your audiences through active, ongoing awareness of your own position and careful communication choices that are attentive to your audience's needs, interests, and questions, as well as to the credibility of your sources. Our approach is grounded in helping you develop ways to credibly participate in the social dynamics of science communication situations across media and genres. The use of AI undermines this approach to science communication by obliterating accurate attribution, risking misinformation, sacrificing the human voice, and dodging the important work of learning about your audience and building your relationship with them. We believe the genre- and rhetorically aware approach this book is anchored in is more important than ever for navigating a science communication landscape that has been refigured by the COVID-19 pandemic. For these reasons, while you may encounter very specific applications of AI tools for the production of some genres (for example, tools for sound-editing a podcast), you should not expect robust engagement with AI from the chapters in this book. We don't offer strategies for interacting with AI in this textbook, except insofar as it's a quagmire to be avoided.

Yet we argue that shifting ground has always been a feature of public science communication. In fact, one might argue that the post-WWII academic research-media complex that privileged peer-reviewed academic journals and provided relatively stable, predictable institutional pathways for the dissemination of research was an anomaly in scientific discourse rather than the norm. The difference today is the speed at which the media ecology and genre landscape evolve. Public science's ground state is an unruly paradox, where we always find ourselves sifting through an infodemic of findings that are of variable quality and complex intent in the study of a universe that never reveals its secrets easily and always throws up surprises.

HOW TO USE THIS BOOK

Amid this fast-changing ecology, our volume aspires to provide not a rule book, nor a set of templates and formulas, but rather a collection of rhetorical strategies and tools for thinking across science communication genres with the expectation that they will change—and that they are part of change. This book is designed as a guide for communicating science in public for scientists or science communicators who want to deepen their understanding of how to engage in compelling ways with audiences who have varied levels of scientific expertise. The book as a whole will equip you with core strategies and tools for communicating science in public and give you opportunities to explore and try your hand at specific, common genres.

This book is divided into two parts. The chapters in Part One, including this introductory chapter, provide a foundation for communicating science in public across genres by introducing key concepts, rhetorical tools (Chapter 2), and principles for inclusive and accessible science communication (Chapter 3) that will be useful to you no matter what genres you plan to communicate in. We suggest that you begin by reading the chapters in Part One to ground yourself in the rhetorically aware, genre-based approach we use throughout this book, and to equip yourself with tools to communicate in ways that engage your audience whether you choose to create a profile or podcast, op-ed or science exhibit, or any other genre or kind of communication, whether we cover it in the second part of the book or not. Equipped with these overarching strategies and principles from Part One, you'll be ready to turn to Part Two of the book and begin learning about and working in a specific genre.

Part Two of this book is comprised of genre-focused chapters; each of these chapters focuses on a different way you might communicate science in public. While they cover varied genres—everything from op-eds, to personal narratives, to podcasts and museum exhibits—these chapters follow a similar structure to teach you about the genre and get you started thinking about it for yourself. These chapters each offer histories and examples of the genre, key considerations for navigating the affordances and constraints, or strengths and limitations, of a given genre, and a process for composing, including steps for brainstorming, composing, revision, reflection, and ideation.

You can explore the chapters in Part Two in any order. You might let a sense of the audience you want to reach guide you. Do they listen to podcasts (Chapter 8)? Or are they more likely to engage with a science exhibit (Chapter 5)? Alternatively, the content might determine which genre you decide to explore first. Does the idea you want to communicate feature a specific scientist who could be featured in a profile (Chapter 7)? Do you have a personal stake in, or experience related to, the research that might make it a good opportunity for a personal narrative (Chapter 4)? While some readers might know exactly what they want to make and head straight to that chapter, we anticipate that others might benefit from exploring a few chapters to learn about the opportunities they offer and the audiences they can help you reach.

WHY A GENRE APPROACH TO SCIENCE COMMUNICATION?

The choice to structure this guide to communicating science in public by genre is based on the way we have been teaching undergraduates in our science communication program at Stanford since 2013. One reason we structure our program in this way is to give students ample opportunities to experiment with different ways of communicating, to make varied kinds of communication artifacts that reach varied audiences, and to figure out the genres or combination of genres that help them communicate the science they care about to the audiences they are invested in reaching. But there are also several other rationales and benefits to this approach, and understanding these can help you get the most out of the genre-based chapters.

Understand Science Communication as Social Engagement with Dynamic Audiences

Genres are useful as a way to classify different forms of communication. Genres you probably engage with every day include emails, text messages, and research articles. Genres for public science communication covered in this book include podcasts, opinion articles, and personal narratives. But what makes a specific genre a genre? What makes a science opinion article, for example, different from a science profile? And what makes it effective?

In her highly cited article "Genre as Social Action," writing studies scholar Carolyn Miller defines genre as "typified rhetorical ways of acting in recurring situations" (1984, p. 151). According to Miller, a genre is an act of communication that behaves in a stable or typical way in a repeated social situation. Miller's definition helps us see that, in order to really understand what makes a genre cohere into a specific, classifiable thing and learn how to work within it, we have to see genres as already socially situated and participating in what genre studies and rhetoric of science scholar Charles Bazerman calls "complex forms of social organization" (2000, p. 6). Throughout this book, as you learn about different genres and strategies for communicating science effectively in public, you will notice that we continuously draw your attention to the social situation you are participating in. If you are designing a museum exhibit, for example, where will it be located and what visitors do you anticipate? If you are writing an op-ed, what do your readers already know about the topic or issue? What questions and concerns do they have? Where and how will they be reading the op-ed? From invention (coming up with your initial ideas) through composing, revision, and reflection, we invite you to think about the complex needs, interests, beliefs, and background knowledge of your audiences, where your audience will encounter what you are creating, and what they are expecting from the genre and from you.

Learning a genre means learning to see communication as community participation rather than just information transfer; it requires that you honor your ethical obligations and consider your positionality vis-a-vis your audience, something we address in more detail in Chapter 3. This approach makes what rhetoric and composition scholar Anis Bawarshi calls "genred discursive spaces" visible to science communicators. Our goal is, again drawing from Bawarshi's work on genre and invention, to "foster [communicators'] critical awareness of what genres help us do and not do … enabling them to participate in these spaces more meaningfully and critically" (2003, p. 18).

A common misconception about public-facing science communication is that it is a one-way act of translation of a technical text—such as a research paper meant for other specialists in a scientific field—into a form that will be more accessible for a generic "general audience." We find that viewing science communication in this way promotes a deficit model of science communication, where our audience is viewed as a kind of empty vessel needing to be "filled up" with scientific knowledge. While removing jargon or finding simpler ways to describe complex processes can be an important part of science communication, it is only one part of the work. To build reciprocal trust, we need to learn to conceive of our audiences in more nuanced, complex, and active ways. A genre approach helps us think through how to actively engage audiences that have their own interests, preferences, questions, and agency in science communication situations.

As you experiment with different genres in our genre-based chapters, we hope this approach will help you avoid an oversimplified, one-way story about science communication. Instead, we hope you will understand how a given genre and its audience are also acting on you in ways that may be both valuable and constraining. By working within a genre approach, you will learn that science communication is not the process of taking fully formed scientific ideas and delivering them to an audience; instead, you will experience the ways that communication can shape and form your own ideas.

Genre and Invention

Working in careful ways within different genres also opens up opportunities in terms of invention and discovery—it can help you find something to say. Bawarshi's work on genre and invention reminds us that genres act on writers and helps us see genres as part of what can provoke a writer to write, giving them both occasion and motivation.

As science communication teachers, we see this play out in our courses on individual genres, where careful study of the genre fosters motivation in our students to try the genre for themselves. The genre-focused chapters of this book are designed to replicate this experience: detailed exploration of the history of each genre and of rich examples leads into invention exercises to support your own work within the genre.

We also hope you notice how invention strategies from one genre are portable to other genres and media. For example, while you may not be interested in designing an interactive science exhibit (Chapter 5), some of the exercises in that chapter may be useful for working with another genre. That chapter's exercise, which asks you to anticipate the experience of a visitor who engages for 30 seconds, 5 minutes, or a half hour with your exhibit, can be easily adapted to genres that are usually written, like the personal narrative or opinion piece. What would someone who glanced at your op-ed on their phone take away? What about someone who skimmed quickly? What about a slow and careful reader? Invention, composing, and revision exercises throughout the book in particular genres will also offer you ways of thinking about communication that can be generative across genres.

While generative AI tools can promise to mimic a certain genre in terms of its moves or common features, we will be asking you to consider the needs, interests, and knowledge of your audience and consider how those factors can or should shape your choices as a communicator. Engaging in that process is what makes you a communicator who can build and sustain trust and an active relationship with your audiences: something that AI can never do.

Become a Nimble Communicator across Genres

As the portability of these invention strategies invites us to consider, the genres in this book are not stable, separate things. Genres have never been completely distinct from one another, and they are increasingly blurring and being rearranged, especially in the present digital communication landscape (and especially in the context of the infodemic described above). A print piece is often simultaneously an electronic one, for example, even though print and electronic media engage readers in very different ways. A podcast might integrate science journalism with sound art, storytelling, and philosophy. Effective science communicators need to anticipate the permeable boundaries between genres and be nimble in moving across them.

The blurring of genres is sometimes a risk to pay attention to: what happens, for example, if a popular publication misrepresents your research article in the interest of a catchy but misleading lede or headline? But it can also be an opportunity for communicators to engage different audiences and thereby develop their ideas. For example, writing a concise museum exhibit label might help you think about how to concisely tell the story of your research in a media interview. Experimenting with different genres gives you the opportunity to extend and adapt your ideas to participate in a different rhetorical situation with different, repeated patterns and conventions.

IDENTIFYING WHAT YOU WANT TO COMMUNICATE

One thing this book will not do is tell you what scientific ideas or questions to communicate. This book assumes that you have a scientific topic or question you want to communicate to new audiences. For some of you, we know the topic that interests you may itself have come from the public: their concerns or curiosity may drive your research.

We recommend that you start any project by familiarizing yourself with some of the most prominent research literature on the topic: the studies that are most cited or those that make an important contribution. If time permits, consider drafting a literature review, a genre that scientists write for each other. Literature reviews provide a comprehensive summary that identifies a research area and synthesizes the scholarship on your chosen topic of interest. Writing a literature review will help you become comfortable with the scientific vocabulary and discern the patterns and trends in the research conversation: how does one paper build on the next? Where is there agreement? Disagreement? What have researchers overlooked? If you're tight on time, consider composing an annotated bibliography, collecting the major studies on your topic and then writing single paragraphs for each that tightly summarize the purpose, methods, findings, and limitations of the studies. These genres—literature reviews and annotated bibliographies—are often read by non-specialists, especially scientists not in the exact field who are looking for an introduction to a research area, but they are not typically written for or read by a wider public. When you have a clear sense of a topic, this book can help you think through how to communicate what you've learned in the most creative, audience-aware way possible, taking into account which parts of your research are most relevant or timely for a public audience.

In an ongoing infodemic—a media ecology that is awash in bad information—we invite you to develop your skills to communicate science carefully and strategically, in dialogue with a public that is eager to learn with you.

REFERENCES

Ananya (2025, September 23). AI models are using material from retracted scientific papers. *MIT Technology Review*. https://www.technologyreview.com/2025/09/23/1123897/ai-models-are-using-material-from-retracted-scientific-papers/

Andersen, R. (2026, January 24). Science is drowning in AI slop. *The Atlantic*. https://www.theatlantic.com/science/2026/01/ai-slop-science-publishing/685704/

Bawarshi, A. S. (2003). *Genre and the invention of the writer: Reconsidering the place of invention in composition*. Utah State University Press.

Bazerman, C. (2000). A Rhetoric for literate society: The tension between expanding practices and restricted theories. In M. D. Goggin (Ed.), *Inventing a discipline: Rhetoric scholarship in honor of Richard E. Young* (pp. 5–28). National Council of Teachers of English.

Bitzer, L. F. (1968). The rhetorical situation. *Philosophy and Rhetoric*, *25*, 1–14. https://www.jstor.org/stable/40237697

Csiszar, A. (2024). Blurry authorship: Originality in science before and after large language models. *Historical Studies in the Natural Sciences*, *54*(5), 611–616.

Ghebreyesus, T. A. (2020, February 15). WHO Director-General address, *Munich Security Conference*. World Health Organization. https://www.who.int/news-room/speeches/item/munich-security-conference

Kornbluh, A. (2024) *Immediacy: Or, the style of too late capitalism*. Verso.

Miller, C. R. (1984). Genre as social action. *Quarterly Journal of Speech*, *70*(2), 151–167.

Postman, N. (2000, June 16). The humanism of media ecology [keynote address]. Media Ecology Association Convention, Fordham University, New York. https://www.media-ecology.net/publications/MEA_proceedings/v1/postman01.pdf

Prasad, V., & Flier, J. S. (2020, May 14). Covid-19, a 'supernova in human history,' will need multiple perspectives to understand and manage. *STAT*. https://www.statnews.com/2020/05/14/underderstanding-covid-19-supernova-human-history-multiple-perspectives/

Rini, R. (2020). Deepfakes and the epistemic backstop. *Philosopher's Imprint*, *20*(24), 1–16. https://philarchive.org/rec/RINDAT

Tyson, A., & B. Kennedy (2024, November 14). Public trust in scientists and views on their role in policymaking. Pew Research Center. https://www.pewresearch.org/science/2024/11/14/public-trust-in-scientists-and-views-on-their-role-in-policymaking/

Wells, G. (2020, May 15). Doctors are tweeting about coronavirus to make facts go viral. *The Wall Street Journal*. https://www.wsj.com/articles/doctors-are-tweeting-about-coronavirus-to-make-facts-go-viral-11589558880

Whitaker, M., Rodrigues, S., Cooke, G., Virlon, B., Donnelly, C. A., Ward, H., Elliott, P., & Chadeau-Hyam, M. (2025). How COVID-19 affected academic publishing: A 3-year study of 17 million research papers. *International Journal of Epidemiology*, *54*(3). https://doi.org/10.1093/ije/dyaf058

Zarocostas, J. (2020). How to fight an infodemic. *The Lancet*, *395*(10225), 676. https://www.thelancet.com/journals/lancet/article/PIIS0140-6736(20)30461-X/fulltext

Zuboff, S. (2019). *The age of surveillance capitalism: The fight for a human future at the new frontier of power*. PublicAffairs.

CHAPTER 2

Rhetorical Tools for Engaging Public Audiences across Genres

WHAT YOU'LL LEARN IN THIS CHAPTER

- A rhetorical framework to help you build on your audience's prior knowledge, interests, and communication practices
- Definitions and examples of rhetorical strategies that will engage audiences with science no matter what genre you are composing in: plain style, description, metaphor, storytelling, and voice
- Brief activities throughout to help you use these tools in your own science communication projects

INTRODUCTION

What are rhetorical tools that help us reach public audiences across genres? These are strategies of communication that capture attention, inform, educate, and build trust. You will see these strategies at play across the genre-based chapters that follow and can use them for genres we do and don't cover in this book—even genres that haven't emerged yet! This chapter defines five prominent rhetorical tools in language—plain language, description, metaphor, storytelling, and voice—and illustrates how they work to engage a public audience with science. (For more on

DOI: 10.1201/9781003520887-3

communicating science with visuals, please see Chapter 10 on visualizing data and Chapter 9 on oral communication, which includes a brief discussion of visual arguments in slideware.)

As discussed in the first chapter, science communication has shifted significantly in the last few decades as communicators have moved from a deficit model of audience understanding—the idea that the science communicator is working to fill a lack in the audience—to an engagement model, one that recognizes that the science communicator and the lay audience are in dialogue with each other (Aikin & Sheufele, 2017). Thus, the communication model we propose here is akin to bridge building, where diverse public audiences and the science communicator exchange knowledge. You will need to perform a basic function: make specialist communication, which is often jargon-heavy and presumptive of a great deal of background knowledge, accessible without sacrificing scientific accuracy. Moreover, you will want to invite and respect your audience's attention and anticipate their questions, prior knowledge, beliefs, and communication practices. This can be challenging. The audience may be distracted or confused. The rhetorical tools we review here can help you meet these challenges and others in order to elicit your audience's attention and trust.

In the fourth century BCE, Aristotle famously defined rhetoric as the "faculty of observing in any given case the available means of persuasion" (*Rhetoric*, Book 1, 1991). The implications of this enduring language are worth examining. "Available means of persuasion" points to all the ways we have of persuading one another—everything from visuals, sounds, and gestures to anything that might help you make a point—and their varied availability. "In any given case" emphasizes the importance of context, or what Lloyd Bitzer (1968) called more broadly the "rhetorical situation." Not every communication situation will call for or make available every means of persuasion. To take a simple example, you prepare a slide presentation for a general audience, but when you arrive, your display is not working. How will you pivot to help your audience see the science you care about? The available means are not just material; more often, they are linguistic. For instance, imagine your bioengineering research might become a biomedical device. How will you adjust the way you speak about your findings when you present to investors who are unfamiliar with your research language but interested in funding a new consumer product? That brings us to another important point of Aristotle's: rhetoric requires discernment and skill. That's why, throughout this book, we ask you to pause to assess successful (and unsuccessful) science communication.

This practice builds your faculty for rhetorical analysis and, ultimately, the wisdom to make choices that respect your audience and make the most of the available means of persuasion.

Your audience may change depending on the impact of your rhetorical choices. Science communication is thus a profoundly relational enterprise and, as such, depends on the trust that develops between you and your audience. It is as much an art as a practice. It is also usually a marathon, not a sprint, even with well-honed rhetorical tools; and it always takes much more time and effort to build (or rebuild) trust than it does to break it (Burgoon et al., 2023). We will acknowledge that rhetorical tools can damage trust when a metaphor distorts or a story oversimplifies, for example. When used well, however, to meet the audience's needs at a particular moment in time and place, they can make even the most remote and complicated scientific ideas accessible, interesting, and memorable.

While the focus of this book is on communicating science in public, it's important to remember that scientists navigate many rhetorical situations, and make use of rhetorical tools, in their communication with other scientists all the time. In fact, science communication, and the careful use of rhetorical tools, is crucial to doing science. When quantum theorists needed to describe the invisible inner workings of the atom, they constructed theories using metaphors that compared the invisible atom to visible systems like the solar system in order to describe the arrangement and behavior of subatomic particles. Think, for example, of the electron's *orbit* in the Bohr model of the atom. To take another example, in his 1836 diaries on the voyage of the Beagle, Darwin compares coral to "myriads of tiny architects," a metaphor that eventually gave rise to new and influential geological understandings of how some islands are formed (Darwin Correspondence Project, 2022). Fundamental scientific ideas like a wave of light use metaphors—comparisons between familiar and unfamiliar or invisible phenomena—to explain and predict things in the natural world. In other words, metaphors are part of science (see Bump, 1985, for further discussion). Storytelling is similarly an important rhetorical tool within scientific disciplines that grapple with exceptionally long time scales. As Gary Tomlinson notes in *Culture and the Course of Human Evolution* (2018), because evolutionary biologists and physical anthropologists cannot run experiments to explain their theories, highly descriptive narrative history often plays an important role in how they explain their ideas to one another and to other disciplines. Nasser Zakariya likewise documents how astrophysicists and cosmologists need access to the genre conventions of epic storytelling in order to formulate and explain theories that unfold over the vast timescales of deep time (2019). We note these uses of rhetorical tools in order to underscore the value of developing your rhetorical toolkit not just for communicating science in public, but also for doing science itself.

TOOL 1: PLAIN LANGUAGE

The trend toward ever more specialized knowledge and language in science since the Scientific Revolution has driven the countertrend toward *plain language.* Plain language is simple language used every day by a broad spectrum of people, in contrast to the jargon that a community of specialists has devised. Today, organizations such as the Science Societies, which advocates for and publishes the work of agronomists and soil scientists, recommend "plain language summaries" of members' work. These summaries of journal articles are designed to reach scientists in other fields, undergraduate and graduate students, and readers not fluent in English. To reach these readers, they recommend writing "at a ninth grade level"—what in the United States is the first year of secondary school—and emphasizing research results rather than methods (Science Societies, 2025).

In medicine, plain language has similarly become a crucial communication strategy to ensure patients understand the risks and responsibilities of their care. But as one resource says, "plain language is more than just simple words" (Ridpath et al., 2007). What really makes something "readable" —or accessible—is the overall style and organization of a composition. Both sentences and paragraphs are short so that information is presented in digestible chunks. Plain language in this framework is thus also a sensitivity to the needs of the readers; it means considering carefully what they need to know and in what order. Content may need to be "filtered" (Ridpath et al., 2007). For example, in a review article in the *Review of Geophysics,* the authors state, "We use a Bayesian approach to produce a probability density function (PDF) for S given all the evidence, including tests of robustness to difficult-to-quantify uncertainties and different priors" (Sherwood et al., 2020). Written for geologists and physicists, this sentence makes reference to a number of specialized analytic methods. AGU, the American Geophysical Union (n.d.), suggests translating this sentence into plain language for a broad audience as follows: "In this report we thoroughly assess all lines of evidence including some new developments" (AGU, "Writing"). This plain language summary eliminates all specialized language to emphasize the thoroughness of the research.

We want to emphasize that using plain language is not just about *translating* jargon into more everyday language, which can result in a composition that is just as long and complicated as the original. It is also about *distilling* the scientific ideas to their essence: the core concepts that will help a public audience understand the implications of the findings.

As useful and important as the plain language approach has been to science communication—you'll see us recommending many of its core strategies throughout this book—it may not always be sufficient. Plain language alone may not grab and hold the attention of the audience or help them care about the problem the science is working to address or why it matters now. Nor may it fully anticipate different levels of experience and prior knowledge, and—depending upon context—some audiences may find it condescending. Advocates for plain language in medicine acknowledge that "examples, analogies, and visual aids" may also be necessary to explain more complex concepts to a diverse audience (Ridpath et al., 2007).

Our next chapter will help you think through how to communicate with diverse audiences to ensure your science communication is culturally relevant and inclusive. In this chapter, we build on the important practice of plain language to show how the rhetorical tools of description, metaphor, story, and voice—alone and in combination—can help you recognize your audience's prior knowledge, interests, language practices, and stakes in the research. Moreover, these rhetorical tools will allow you to acknowledge gaps without condescension, scaffold complex concepts for different experience levels, and maintain engagement across varying attention spans.

Effective Plain Language

- Is widely understandable
- Uses everyday words to explain scientific concepts without compromising accuracy
- Breaks content into digestible chunks

Activities to Practice Your Skills in Plain Language

- Pick a scientific topic about which you know a lot: one that you'd be inclined to use a lot of detail and jargon to explain. Call up a family member or friend who knows nothing about it. Try explaining it to them in plain language first in 6 minutes, then in 3, and then in just 1 minute.

- Revision exercise: take a jargon-heavy description from a scientific journal article important to your work and translate it into plain language.

TOOL 2: DESCRIPTION

Vivid description is one of the most powerful tools we have to reach a public audience, rendering processes that are visible to scientists but invisible to most others. When done well, description offers a gateway to new understanding.

Even something as familiar as human aging can become new again when a scientist communicates what they know with fresh description. Atul Gawande is a surgeon, professor of public health at Harvard University, and staff writer for *The New Yorker*. In 2007, he published "The Way We Age Now," a magazine story that shows why geriatrics is a medical subspecialty. But instead of describing all the training of a geriatrician—what medical content they have to master to earn their licenses—Gawande (2007) describes the aging human body, the thing they have to understand closely in order to heal others. If time permits, search for and read Gawande's essay now.

Gawande's opening paragraphs are so striking that readers don't soon forget them. He starts by describing universal human aging, but instead of rendering aging from the outside—what most of us can see in all the elderly people we know and love—he presents it from the inside, from the perspective of a medical doctor, and he starts with an unlikely part of the body, the teeth. Their white enamel, we learn, is the hardest substance in the human body, and yet even it will begin to wear down with age, exposing the darker layers beneath. Blood flow to the jaw lessens, nerves atrophy, and so an elderly person may not even feel the tooth pain that signals a visit to the dentist is in order. The description then turns on a paradox: while the hard parts of our bodies soften, the soft parts harden. A wonderful paragraph describes this process, which Gawande attributes to the redistribution of calcium. The startling consequence, which only he could know as a surgeon: "When you reach inside an elderly patient during surgery, the aorta and other major vessels often feel crunchy under your fingers."

Gawande could have just started his piece with a claim about blood flow to the heart decreasing in the elderly, a fact that is important to the observations he later makes in the piece about the energy levels of patients he is profiling. But that would not have built the same level of understanding in his reader, who instead can now locate this fact within the larger process of aging that Gawande vividly describes. Moreover, the process sets up a perplexing effect that entices the reader to ask about its cause, something that Gawande next addresses as he describes several scientific theories of aging. Finally, the vivid description of universal aging with which Gawande opens his magazine story anchors the individual human stories he tells later—about doctors and patients navigating this process. By building his readers' understanding early through description, Gawande prepares them for the more complex understanding provided toward the end of the story regarding the social implications of aging. By describing a universal process objectively and scientifically early on, Gawande also builds trust with skeptical readers who might not otherwise see the significance of the single human story with which he ends the piece.

How much description is too much description? After Gawande discusses the scientific debate about why we age, he adds more description about how the body ages; not just the universal graying of hair, but also what happens at the cellular level that impacts our skin and organs like our eyes. He can do this because he's writing in the genre of a magazine article, in a magazine known for its long-form journalism. The readers of *The New Yorker* typically have the patience for, and preference for, these expansive discussions. How much description you include of the scientific process will depend on the constraints of the genre you're writing in as well as your audience's attention span. Podcasters often rely on robust descriptions because audiences need to "see" with their ears. But if you're being interviewed for a video news story, for example, or writing a short newspaper op-ed, you will have to keep your description very tight. To meet the expectations of the genre you communicate in, you will need to select details for your descriptions carefully.

Effective Scientific Description

- Assumes the reader is largely unfamiliar with the phenomenon at hand
- Provides value to all readers through fresh description
- Balances technical accuracy with vivid language

Activities to Practice Your Skills in Description

- Read slowly (or play back slowly, if you're listening to) your favorite examples of public-facing science communication. When is description used? How detailed is it? How effective is it at revealing specialist knowledge, and why?
- Start a notebook in which you regularly practice describing what you're seeing, hearing, tasting, and feeling: phenomena of nature and social or digital life—whatever most interests you. What details would you need to share to bring your reader fully into your felt experience? For more on the value of keeping a notebook for your science communication, see Chapter 4 on personal narratives.
- Select a scientific process. It might be a method scientists use to make new knowledge, or it may be something they've observed about our world. Now try to describe it using only sensory details: the colors, shadows, shapes, or lines that can be seen; the sounds that might be heard; the temperatures or textures that might be felt; among other details.

TOOL 3: METAPHOR AND ANALOGY

Description is a literal rendering of a scientific phenomenon, but it may not be sufficient to make a complex concept accessible and memorable. In these cases, science communicators turn to metaphors and analogies to explain what is unfamiliar in terms their audience is familiar with. The metaphor or analogy thereby builds a bridge from old knowledge to new knowledge. A well-chosen metaphor can also show that the science communicator knows their audience and what will appeal to or inform them, strengthening the trust between them.

Aristotle, whom we've already seen was very interested in how to achieve understanding between people, said in his *Poetics* that a metaphor is "giving the thing a name that belongs to something else" (Chapter 21, 1457b, 1987). We do this when we say a clock has a face and hands, for example, parts we typically associate with humans. So it's something that we do in everyday language, often without thinking about it. Yet you might be concerned that giving a scientific phenomenon "a name that belongs to something else" only risks more confusion. You're not wrong—that can happen. But that risk can often be worth taking, as we'll see below, when the scientific concept is very new, very complex, or remote from the audience.

It is not uncommon for people to confuse metaphor with analogy. This is because analogies also work by comparison but tend to be more explicit and systematic than metaphors, often mapping multiple corresponding elements between domains. For example, "the heart is to the body as a pump is to a water system" is an analogy that helps us understand how the human heart works. But because the wording is a little clunky, and concision is often at a premium when you're communicating in public, you are likely to use the more compressed, metaphorical version of this analogy and refer simply to the heart as a pump. In both cases, nevertheless, non-literal language generates insight through juxtaposition. Neuroscience has shown that metaphors based on sensory experiences activate the brain's cortex as well as its language-processing areas, which suggests that we don't process these metaphors as abstractions alone—we also feel them at some level (Lacey et al., 2012)!

Let's look at how this might work in a successful example of science communication. Again in *The New Yorker*, Ted Chiang (2023) worked to explain the computer science behind the game-changing large language model that produced the ChatGPT app. His widely read essay is anchored by a claim that is also an analogy: as his title states, "ChatGPT is a Blurry JPEG of the Web." Chiang was writing soon after the app burst onto the scene in late 2022, and the world was hungry to learn how the technology could generate seemingly magical human-like text. If you haven't already, you might read the essay now.

Chiang begins his magazine story by reminding his readers of how copiers work: by scanning documents and then printing the image file. But what is central to his essay and the analogy is the point that the scan is not total: the file that is printed has almost always been compressed, and not all the information from the original is present in the copy. Put differently, the copying process relies on "lossy compression" to reproduce documents. Over several paragraphs, Chiang's description (see how these rhetorical tools can overlap and interanimate?) builds to the title's analogy with this further explanation:

> ChatGPT retains much of the information on the Web, in the same way that a JPEG retains much of the information of a higher-resolution image, but, if you're looking for an exact sequence of bits, you won't find it; all you will ever get is an approximation.

Chiang then uses the analogy to explain both the internal workings of large language models—their use of interpolation—and some of their effects, such as hallucinations. In this example, something that is more familiar to readers—copiers—is compared with something that was still quite mysterious in 2023—the astonishing output of a popular large language model—to make it more accessible and concrete.

Chiang assumes his audience is conversant with a wide range of technologies, including image file formats such as JPEGs. But Chiang (and his editors) take the risk that his readers will find this older technology more understandable than the new. What's interesting is that much of the essay is about "lossy compression," but because that is as abstract a concept as a large language model, he leads with the idea of the "blurry JPEG" in his title and reiterates it at key moments throughout the essay. Most of the public today can imagine a pixelated, blurry JPEG, which helps them see the seeming lucidity of the chatbot in a new way. If you're someone who can't imagine that JPEG, however, the comparison will not help you understand large language models, and it may even confuse you further. In this case, the analogy obscures rather than illuminates. The metaphors you choose to help you communicate your science need to align with your audience's background knowledge.

Moreover, Chiang's analogy is not neutral: it conveys a point of view. Chiang invites his readers to consider whether, given the choice, they would rather have access to all the documents on the web through a Google search or through their "blurry" paraphrases, what was (and, as of this writing, still is) available through a ChatGPT query. It's clear from the analogy that Chiang would choose the former. Because Chiang has already shown his largely professional readers that he understands them as well as the technology, they are likely to trust this conclusion as well.

Many writers and thinkers would resist the comparison of ChatGPT to a blurry JPEG, including perhaps you. Research shows that students have described the chatbot as a map, a compass, and a bus driver when it helps them organize their research (Jin et al., 2025). These metaphors convey a whole other understanding of the tool. As you develop metaphors in your science communication, you will want to think carefully not only about what they reveal about the scientific process you are communicating but also about their connotations. Try to anticipate whether your audience will be amused, further confused, or even offended by the comparison, and consider whether those reactions suit your rhetorical purpose.

In sum, a metaphor can make a complex or abstract idea memorable so that it can be repeated across audiences. Or it can organize new ideas and observations into a more familiar form to produce new understandings. It may even do real intellectual work that teaches and develops new ideas. In spite of these promising possibilities, many science communication handbooks don't address metaphor, in part because metaphors can be variously interpreted (Bump, 1985). Use them carefully, however, and you will find them integral to your science communication in public. For further discussion of metaphor, see Chapter 6 on scientific opinion writing and Chapter 8 on podcasts.

Effective Scientific Metaphors

- Illuminate rather than obscure
- Take a specific point of view
- Do not reinforce misconceptions
- Are culturally aware
- Help build trust between science communicators and their audience

Activities to Practice Your Skill in Metaphor Writing and Analogy

- Return to some of your favorite popular science communication. When do the writers and speakers turn to metaphor? Which work best? Why?
- Pick a scientific concept that's important to your research or communication agenda and that is also difficult to explain. Generate multiple metaphors or analogies for this concept ("X is like Y because ..."—you might produce 1–8 sentences depending on the complexity of the comparison). Remember that your distinction as a science communicator will often depend on how illuminating and unique your comparisons are.

  Then, for each of the metaphors or analogies you generate, spend some time writing about what they help you say: What can your audience see with them that they couldn't see without them? How might the metaphors obscure or misrepresent the scientific concept? Does the clarity the metaphor brings suit your rhetorical purpose, i.e., what you hope to accomplish with what you are communicating?

You might repeat this process with several scientific concepts.

Pick your two favorite metaphors and share them with a non-expert audience. What do they learn? How might your metaphors be interpreted differently than you intended? What other connotations might your audience bring to the unfamiliar and familiar things you compare in your metaphors? Which is most memorable to them, and why? Has your metaphor achieved its purpose?

TOOL 4: STORY

Stories are a fundamental mode through which humans process information and make meaning of shared experiences (Dahlstrom, 2014). A good story draws on vivid description and apt metaphor and, as we'll see, conveys human voice, too. By "stories," we do not mean fiction or something that's made up; rather, we refer to the structure of narrative, which, even as narrative nonfiction, can evoke feeling.

Western science has expressed skepticism about storytelling, viewing it as a tool that manipulates audiences through emotional appeal or substitutes unrepresentative anecdotes for generalizable data (Weeks, 2024). Yet Indigenous science has always used story to both create and communicate science, and many contemporary science communicators advocate for story as a way to braid knowledge traditions (Archibald, 2008; Atalay, 2020). Green and Doucette (2017), for example, argue that effective science communication requires "uniting science and stories," transforming scientific knowledge into narratives that resonate with public audiences (see also Dahlstrom, 2014; Olson, 2015).

The structure of narrative itself supports comprehension in ways that purely expository writing often cannot. Stories typically follow a narrative arc—a beginning, middle, and end that builds tension through unexpected developments and resolves through change (Joubert et al., 2019). This structure creates momentum that carries readers or listeners through complex material, maintaining their attention while also providing natural opportunities to contextualize methods, explore uncertainties, and explain broader implications. Research demonstrates that stories help audiences understand, process, and recall science-related information more effectively than other presentation formats (ElShafie, 2018). Moreover, when audiences identify with characters in a story whose values resonate with their own, narratives enable them to connect abstract scientific findings to their lived experiences and decisions (Joubert et al., 2019).

Finally, when stories are personal, they often help audiences understand how the science communicator is connected to the topic or concept, building trust with the public as well as the communicator's identity as a science communicator (Alfano et al, 2023).

How can you build your own stories? Stories are made from three components: character, plot, and setting.

- *Character*: An audience needs a person or thing to follow through the narrative. A human character can express feelings such as doubt, curiosity, and excitement, which can make the investigation of science relatable and fun. If you are writing a personal narrative, the main character in the story is you, the writer or speaker, which we explore in some depth in Chapter 4. In some science communication, we should also acknowledge that the main character is not a person but a natural phenomenon such as a comet. But no matter what type of character your story centers on, there is still change over time. That's because a story requires a character who changes themselves or changes their surroundings.
- *Plot*: To create a plot, your character must do something rather than just take in experiences. The plot is the story's sequence of events. In scientific narratives, discovery often drives the plot, as the character moves from confusion or doubt to greater understanding. The character's need for change catalyzes the plot; their ultimate change helps mark the end of the plot.

  The plot should be organized into scenes in which the character does something ("showing" or action), followed by a moment of reflection that tells us what it means ("telling" or reflection) (Figure 2.1). This is what's known as the "layer cake" model of storytelling. Sometimes the action in a single scene of scientific storytelling is your description (see our discussion of Tool 1 above) of a phenomenon out there in the world. Oscillate between showing and telling as your character changes over the course of the plot.
- *Setting*: All stories take place in a specific time and place. Think: a specific lab, a specific doctor's office, a specific field site at a precise moment in time and thus within the history of science. This context has a significant effect on the story. Indeed, most stories unfold in a way that could only happen in their specific setting. Setting is so

FIGURE 2.1 In a story, each scene is composed of a character performing an action followed by some reflection. In this figure, five scenes create a plot.

> important that even though it's third on this list, when you sit down to compose your own story, you might consider starting with setting. It's usually responsible for a good deal of what's known as the "inciting incident," the moment or event that catalyzes the action and the need or desire of your main character. Put differently, the setting often sets up the problem that your character needs to resolve.

Let's see how these story components are used in a podcast, RadioLab's 2015 episode "Staph Retreat" (Nasser, 2015). If you don't know it already, give it a listen! RadioLab is a podcast that's well known for making complex ideas accessible and for its artful layering of expert perspectives, anecdotes, and relevant background sounds.

In "Staph Retreat," the setting is very important and made explicit: initially, the podcast takes us to 1928 London and the lab of Alexander Fleming, who happens to be growing *Staphylococcus* bacteria. At this time, there is no known cure for a staph infection, which, as the science writer they're interviewing reminds us, is caused by bacteria that live on much of our body and can become deadly if they make their way into our bodies via a cut, for example. In this lab, a window is left open while Fleming is away on vacation, and when he returns, instead of Petri dishes filled with flourishing bacteria, as one would expect, he finds dead zones. This is the inciting incident that invites the question "why?"—and it could only happen in the time and place of this particular setting.

The plot of the podcast follows the powerful "what happened → then realization" structure in a series of scenes. In the first scene, as we've seen, Fleming leaves the window open and realizes that a mold is secreting a compound that is killing the bacteria around it. Narrative and scientific explanations are intertwined. As the guest explains, "Infectious diseases that had been killing people for as long as we had been people suddenly could be stopped." The story would end there, but a new problem emerges quite quickly—antibiotic-resistant bacteria—which drives the story forward.

The individual scenes are structured chronologically. That's because stories typically unfold in a linear fashion over time. But what makes this episode of the podcast distinctive is that it goes back in time even as science is moving forward. The microbiologist finds an unlikely partner in a historian, specifically a specialist in Viking Studies who has located a promising medieval handbook of healing. Together, they test one of the medieval "recipes" and find that it successfully suppresses a bacterial infection. Old cures, it turns out, may beat new bacteria. The overall plot of this podcast thus moves through several scenes of discovery and setback before finally ending with a more comprehensive understanding of antibiotic resistance. Suspense—will the cures work? Why do they work?—keeps the audience engaged throughout.

The cast of characters in this podcast is eclectic and rather extensive. The voices of several podcast hosts, a reporter and producer from the RadioLab team, a science writer, and several professors are all featured. The interaction reinforces that science is a social enterprise. But there is one set of characters the audience can consistently identify with: the hosts. They actively process information together on mic, asking questions, expressing wonder or confusion, and essentially learning alongside the listener rather than just lecturing. The other main characters we follow in this podcast are the bacteria, whose resilience humans must continuously confront and overcome.

This particular podcast is in large part about serendipity, the perhaps surprising importance of dumb luck to the making of science. In this case, story is an especially helpful tool for helping us understand how scientific knowledge evolves, but especially when embedded in a popular podcast, it does not tell us much about the mechanics of antibiotic resistance or even how antibiotics work to suppress bacterial infection. We just know that they do after listening to this episode. The point here is that all stories are selective, not totalizing, and thus cannot serve all communication purposes.

You will want to think carefully about whether and how story can serve your own rhetorical goals. And you will want to be especially careful about the conclusions you draw from story. If you are arguing that your story is representative of a larger trend, then you will want to show your audience why you know it is.

"Small stories" or anecdotes can be just as powerful as extended narratives. They are often used as entry points to or illustrations of larger concepts. See our discussions in the chapters on op-eds and museum exhibits, for example, for ways to use anecdotes in those genres (Chapters 5 and 6). For more on story arcs and character development, see the chapters on the profile and personal narrative (Chapters 4 and 7).

HERE ARE SOME STORY ARCS THAT CAN BE USED ALONE AND IN COMBINATION FOR SCIENCE COMMUNICATION:

THREE ACTS

1. A description of the world before the scientific discovery
2. A description of the scientific discovery process
3. A description of how the world changes as a result

RISING TO THE CHALLENGE

- A character discovers and accepts a new challenge
- The journey toward achieving this goal (or learning from setbacks)

CHAOS TO MEANING

- Confusing or disorienting setbacks in research or understanding
- Movement from uncertainty to greater certainty
- Resolution through reflection, analysis, and action

CHANGING THE WORLD

- Success stories showing how science improves understanding or conditions
- Stories that balance optimism about scientific progress with a realistic depiction of complexity

Effective Scientific Storytelling

- Follows a character who changes themselves or who changes the world in a specific time and place
- Layers narrative with explanation of the science
- Avoids oversimplification of scientific impact

Activities to Improve Your Storytelling in Science Communication

- Read Atul Gawande's "The Way We Age Now" in *The New Yorker*. Consider printing it out. Color-code or annotate the showing versus telling components. Does Gawande use the layer cake approach to scientific storytelling? How does he transition between the two components of story? Identify how Gawande weaves scientific information into narrative structure.
- Tell a story out loud to a friend or family member about your scientific research, with you as the main character. Focus on engaging your audience using plot in a specific setting. Then incorporate feedback from your listener. When was your audience most engaged? When did they need more information from you? Finally, repeat the process, only retell the same story from another character's perspective (e.g., research director, research subject, or policymaker).
    - *Reflection*: What do these stories help your audience see about the science? What do they obscure? What do you have to leave out? Is story an appropriate rhetorical tool for communicating your science?

TOOL 5: VOICE

Voice is our final tool. Drawing on the other four tools, it is distinctly human and thus the most important. Voice includes the scientific content that you choose to communicate, the stance you take on it, *and* how you present it, down to the individual words you choose. Your word choice and sentence structure, and the design of your graphics, also comprise your style, which is often used interchangeably with voice. In our discussion here, we'll think of style, particularly your verbal style, as an important part of your voice but not all of it.

While you have "a voice" that you can use to teach and make change, what we are suggesting here is that your voice is plural: you can adapt it to suit the rhetorical situations in which you find yourself communicating science. You will have to make choices within the genre constraints we discuss in the following chapters, and you will also want to modulate your voice—its tone and pace, for example—based on what you learn about your audience, sometimes, if the audience is live, even as you are in the midst of communicating. As bell hooks argues in *Teaching to Transgress*, "The engaged voice must never be fixed or absolute but always changing, always evolving in dialogue with the world beyond itself" (hooks, 2014).

You might feel that such a notion suggests we are asking you to communicate inauthentically—and that having multiple voices might actually undermine your audience's trust in you. Instead, however, we're asking you to consider how these voices are all reflections of you, what you know, and what you want to say, and that you're adapting your language and content to your audience's needs.

For example, you may have been taught not to use personal pronouns in your science writing. The first person, especially, has been taboo in scientific communication in favor of a seemingly more objective style. But both theoretical (see Donna Haraway's essay "Situated knowledges" [1988]) and empirical research (see Helen Sword, 2012) suggest that the first person is not only used in scientific writing but, when combined with what Haraway calls critical positioning, is also the most truthful. You'll want to consider carefully when and how you use the first person. There will, of course, be occasions and genres where you will want to write in a more objective style and avoid the first person, in what Peter Elbow calls "author-evacuated prose" (Elbow, 1991). That's at the most objective end of the spectrum; for the most subjective, see Chapter 4 on personal narratives. There, we discuss a first-person voice that is so vivid and direct that you feel as though you're in the presence of the writer.

Our conception of voice draws on its oral dimensions—that each human, when they talk, produces a unique blend of sound influenced by their anatomy and speech patterns (Davidson, 2024). When you write with voice, your audience can hear you and feels they are in the company of a distinctive human, not bot-generated text. In this era of ubiquitous AI slop, voice in your science communication may be the most vital rhetorical tool at your disposal.

Let's reflect on some of the key components that make up our voices. Like our other rhetorical tools, these can overlap and work in concert with one another:

- *Diction*: Our word choice reflects both our expertise and our cultural backgrounds. On the one hand, you will want to choose words that reflect the genre conventions of the form you are composing in and, on the other, you will want them to be somewhat fresh so that you help your audience see something new, per our discussion of description above. When communicating in public, you will want to balance technical precision with plain language to convey awareness of your audience's needs. In the next chapter, as we discuss inclusive science communication, you will also become more aware of the cultural and contextual considerations of your communication, which will impact your word choice.
- *Syntax*: Our sentence structure reflects both our own logic and our attention to what our audience needs from us. Our sentences can be short or long, grammatically complex or quite straightforward. Syntax can convey both your personality—perhaps you consider yourself a straightforward person who communicates concisely no matter what—and your expertise, as reading and listening to a lot of research conversations may have made you comfortable with long sentences, which researchers often favor.
- *Pace*: A significant component of your voice is the speed with which you deliver information. In part, your pace will be driven by the constraints of your genre: if you are writing an opinion piece for a newspaper, for example, you're likely to need to work within a very limited word count and move through your information and argument quickly. Still, you can make choices about what to focus on and use pace to build tension in a story or provide an extended reflection. Pace impacts the other rhetorical tools discussed in this chapter, too. Your metaphors may be very compact or your analogies drawn out; you may choose to tell a brief anecdote or spin a long story. You may need to limit these kinds of rhetorical flourishes, which is often a good thing, so that those you do use shine. You should work to match your pace to the audience's attention patterns and to honor the constraints of the genre you're composing in strategically.

- *Tone*: Your attitude toward your subject is communicated through your tone. Much scientific communication favors an objective tone to establish facts and even to recommend policy based on data. However, you may also choose to convey optimism, skepticism, or urgency, among other feelings about your subject, which is something we often do through our word choice, volume, pitch, and pace. When communicating in public, you will want to establish an appropriate emotional register based on your rhetorical purpose and genre. In an oral presentation to a general audience, for example, you may be able to use more appeals to emotion, such as humor. As you consider what tone to strike, you will want to balance your authority with your approachability.

Let's look at an example of how a writer conveys voice on the page. Robert Sapolsky is a professor of biology, neurology, and neurosurgery at Stanford University. He has taken a keen interest in communicating with the public, penning, among other works, crossover books such as *Why Zebras Don't Get Ulcers* and *The Trouble with Testosterone*. Most recently, he published the *New York Times* bestseller *Determined: A Science of Life without Free Will* (Sapolsky, 2024). Penguin, a popular press, published this 500-page book with both endnotes (for technical references) and footnotes (for clarification—a voice from the margins). Like all Sapolsky's books, *Determined* works to share an astonishing amount of research with a broad audience. It demonstrates well why voice is important to much more than communicating personal topics, and why we might want to think of voice as comprehensive style.

What's striking about Sapolsky's prose is that you can hear his voice as you're reading: you feel as though he's talking to you. He uses the first-person liberally and opens with an anecdote from his college days. He makes references to a number of disciplines–from anthropology to history to neuroscience and physics—while weaving in a quote from a hit play (a footnote explains that his wife directs musical theater). His tone varies, moving from invitational to funny to earnest and urgent; his own emotion is expressed on the page. Pace varies, too, as he balances breathtakingly quick summaries of his own work—one 500-page book is described in just a few paragraphs—with deeper dives into technical topics such as chaos theory, complete with graphics. He often uses direct address, anticipating his audience's questions, and interjects to anticipate their responses to the science he's sharing. Weird. Obviously. Awful. These one-word, highly conversational sentences punctuate much longer, formal sentences, some up to five typescript lines long.

In this brief discussion, we suggest that voice is important to making expertise accessible *and* engaging. The rhetorical tools we discuss in this chapter work together to achieve Sapolsky's purpose of convincing you of something he knows most of his audience (including us!) will resist: that we have no free will and that science can prove it. Because he's writing a trade book, his genre is highly flexible and capacious. He has the space to develop and express his voice by articulating a controversial point of view, ranging widely to support it, and modulating his style in diverse ways in the interest of keeping his audience engaged. One fascinating feature of the paperback edition of this book that we want to call out: it has an epilogue, "Things I wish I'd Written in the Hardcover Edition" (Sapolsky, 2024). It's a wonderful example of bell hooks' admonition to evolve your voice in dialogue with the world around you. In the new epilogue, Sapolsky engages with his critics, acknowledges the book's limitations, and works to clarify still-murky answers in only the way he could. Put differently, his voice is distinctive, responsible, and responsive to his public.

Given the proliferation of unvetted, AI-generated content, today more than ever, an authentic voice is crucial to science communication. A human voice in communication is irreplaceable: unlike AI, it can explain its reasoning, express vulnerability, and take responsibility for its errors. As it turns out, generative AI has made it even more important that we demonstrate that we are human in our science communication. One of the surest ways to do that is to develop a distinctive voice. Our voices reflect our humanity, facilitating real connection and trust with our audience.

You never have to find a voice, as you always already have one, but you can strengthen it through practice. To develop as a science communicator, we encourage you to move beyond imitation of others' styles—though that can be a good place to start—to develop authentic expression. The danger is that you become formulaic in your adherence to genre conventions. This book instead encourages you to develop more versatility and thus confidence in meeting the moment with the version of your voice to which your audience might best respond. You can use your voice with *discernment*, sometimes ceding it very deliberately and, at other times, pumping up the volume on your own particular style and point of view.

Effective Use of Voice in Scientific Communication

- Draws on all the available means of persuasion strategically
- Adapts to genre conventions and audience needs
- Maintains a distinctive style that builds trust
- Reflects our humanity: our fallibility and capacity for learning

Activities to Enhance Your Voice in Science Communication

- Read your favorite popular science communication for voice. Which genres tend to have the most distinctive voices? Which tend to adopt a more objective style? Why?
- Susan Orlean (2007), a well-known non-fiction writer, suggests the following technique for voice development in writing: Read your work aloud to hear how you tell stories on the page. Ask yourself: Does it sound real? Would I have said it that way? Really listen to your work to refine an authentic voice.
- Take a "voice inventory." Consider all the ways you speak and write: at home, in more professional settings, and online. How might you authentically integrate aspects of these voices into science communication? (For more on this exercise, see Chapter 4 on personal narratives.)

CONCLUSION

Throughout this chapter we have shown how five rhetorical tools—plain language, description, metaphor, storytelling, and voice—work synergistically to communicate science to a broad audience in diverse genres, from magazine stories to podcasts. If you practice these tools alone and in combination, you will find that you are able to draft more confidently in new genres because you have the foundational skills that you can then adapt to the particular genre conventions and rhetorical situation. Moreover, these tools, having already endured across millennia of human communication, will no doubt continue to find purchase even in an ongoing infodemic and in emerging forms of communication. Perhaps more importantly, as you develop your confidence in using these tools in credible ways, you will develop a voice that is authentic to you and that builds trust not only in your communication but also in the vital science you are inviting your audience to consider.

FURTHER RESOURCES

American Geophysical Union. 2026. "Plain Language Summaries." https://www.agu.org/publications/authors/journals/plain-language-summary#summarize-science#summarize-science

Tips for explaining science in short and simple terms with links to further resources.

Centers for Disease Control and Prevention. United States Centers for Disease Control Clear Communication Index. May 29, 2025. https://www.cdc.gov/ccindex/index.html

A research-validated tool that helps writers develop and assess public communication. It includes specific metrics for clarity and includes health-/science-specific guidance.

Green, S. J., Grorud-Colvert, K., & Mannix, H. (2018). Uniting science and stories: Perspectives on the value of storytelling for communicating science. *Facets*, *3*(1), 164–173. https://doi.org/10.1139/facets-2016-0079

A scholarly resource that includes helpful tips for becoming a better storyteller.

REFERENCES

Aikin, H., & Sheufele, D. (2017). Overview of the science of science communication. In K. H. Jamieson, D. Kahan, & D. Sheufele (Eds.), *Oxford handbook of the science of science communication* (pp. 25–34). Oxford University Press.

Alfano, C., Polk, E., & Stonaker, J. (2023). Science communication across disciplines:Reflecting on STEM identity building through notation in science communication eportfolios. *Across the Disciplines*, *20*, 135–156. https://wacclearinghouse.org/docs/atd/volume20/alfanoetal.pdf

American Geophysical Union. (n.d.). *Writing an effective plain language summary*. https://www.agu.org/publications/authors/journals/-/media/556ae47136464f2cbf448f087d8d11a6.ashx

Archibald, J. A. (2008). *Indigenous storywork: Educating the heart, mind, body, and spirit*. University of British Columbia Press.

Aristotle. (1987). *Poetics* (S. Halliwell, Trans.). University of North Carolina Press.

Aristotle. (1991). *On rhetoric: A theory of civic discourse* (G. A. Kennedy, Trans.). Oxford University Press.

Atalay, S. (2020). Indigenous science for a world in crisis. *Public Archaeology*, *19*(1–4), 37–52. https://doi.org/10.1080/14655187.2020.1781492

Bitzer, L. F. (1968). The rhetorical situation. *Philosophy & Rhetoric*, *1*(1), 1–14. https://www.jstor.org/stable/40236733

Burgoon, J., Dunbar, N., Metzger, M., Staphopoulis, A., Metaxas, D., & Nunamaker, J. (2023). The psychology of trust from relational messages. In Martha Peaslee Levine (Ed.), *The Psychology of Trust* (pp. 57–74). IntechOpen. https://doi.org/10.5772/intechopen.101182

Bump, J. (1985). Metaphor, creativity, and technical writing. *College Composition & Communication, 36*(4), 444–453. https://doi-org.stanford.idm.oclc.org/10.2307/357863

Chiang, T. (2023, February 9). ChatGPT is a blurry JPEG of the web. *The New Yorker.* https://www.newyorker.com/tech/annals-of-technology/chatgpt-is-a-blurry-jpeg-of-the-web

Dahlstrom, M. F. (2014). Using narratives and storytelling to communicate science with nonexpert audiences. *Proceedings of the National Academy of Sciences, 111* (supplement_4), 13614–13620. https://www.jstor.org/stable/43043099

Darwin Correspondence Project. (2022). *Darwin and coral reefs.* University of Cambridge. Retrieved January 19, 2026, from https://www.darwinproject.ac.uk/commentary/geology/darwin-coral-reefs

Davidson, M. (2024). Voiceprint analyses. Ebsco research starters. https://www.ebsco.com/research-starters/applied-sciences/voiceprint-analyses

Elbow, P. (1991) Reflections on academic discourse: How it relates to freshmen and colleagues. *College English, 53*(2), 135–155. https://doi-org.stanford.idm.oclc.org/10.2307/378193

ElShafie, S. J. (2018). Making science meaningful for broad audiences through stories. *Integrative and Comparative Biology, 58*(6), 1213–1223. https://doi.org/10.1093/icb/icy103

Gawande, A. (2007, April 30). The way we age now. *The New Yorker.* https://www.newyorker.com/magazine/2007/04/30/the-way-we-age-now

Green, S. E., & Doucette, D. (2017). Uniting science and stories: Perspectives on the value of storytelling for communicating science. *FACETS, 2*(2), 1015–1022. https://doi.org/10.1139/facets-2016-0079

Haraway, D. (1988). Situated knowledges: The science question in feminism and the privilege of partial perspective. *Feminist Studies, 14*(3), 575–599. https://doi-org.stanford.idm.oclc.org/10.2307/3178066

hooks, b. (2014). *Teaching to transgress.* Routledge.

Jin, F., Sun, L., Pan, Y., & Lin, C. H. (2025). High heels, compass, spider-man, or drug? Metaphor analysis of generative artificial intelligence in academic writing. *Computers & Education,* 228. https://doi.org/10.1016/j.compedu.2025.105248

Joubert, M., Davis, L., & Metcalfe, J. (2019). Storytelling: The soul of science communication. *Journal of Science Communication, 18*(5), Article E. https://doi.org/10.22323/2.18050501

Lacey, S., Stilla, R., & Sathian, K. (2012). Metaphorically feeling: Comprehending textural metaphors activates somatosensory cortex. *Brain and Language, 120*(3), 416–421. https://doi.org/10.1016/j.bandl.2011.12.016

Nasser, L. (Host). (2015, March 20). Staph retreat (No. 4) [Audio podcast episode]. In M. Kielty & S. Wheeler (Executive Producers), *Radiolab.* WNYC Studios. https://radiolab.org/podcast/best-medicine

Olson, R. (2015). *Houston, we have a narrative: Why science needs story*. University of Chicago Press.

Orlean, S. (2007). On voice. In M. Kramer & W. Call (Eds.), *Telling true stories: A nonfiction writers' guide from the Nieman Foundation at Harvard University* (pp. 158–161). Penguin.

Ridpath, J. R., Greene, S. M., & Wiese, C. J. (2007). *PRISM readability toolkit* (3rd ed.). Group Health Research Institute. https://prism.kpwashingtonresearch.org/documents/PRISMReadabilityToolkit_ThirdEdv6_062210.pdf

Sapolsky, R. M. (2024) *Determined: A science of life without free will*. Penguin.

Science Societies. (2025). *Plain language summary guidelines*. https://www.science-societies.org/publications/journals/plain-language-summaries

Sherwood, S. C., Webb, M. J., Annan, J. D., Armour, K. C., Forster, P. M., Hargreaves, J. C., et al. (2020). An assessment of Earth's climate sensitivity using multiple lines of evidence. *Reviews of Geophysics*, *58*, e2019RG000678. https://doi.org/10.1029/2019RG000678

Sword, H. (2012). *Stylish academic writing*. Harvard University Press.

Tomlinson, G. (2018). *Culture and the course of human evolution*. University of Chicago Press.

Weeks, W. B. (2024). Storytelling in scientific conferences: Mitigating misinformation risk. *International Journal of Public Health*, *69*, 1607289. https://doi.org/10.3389/ijph.2024.1607289

Zakariya, N. (2019). *A final story: Science, myth, and beginnings*. University of Chicago Press.

CHAPTER 3

Making Science Communication Inclusive and Accessible

WHAT YOU'LL LEARN IN THIS CHAPTER

- A definition of inclusive science communication (ISC) as both an applied practice and a social movement aimed at actively redressing and healing past harms
- A discussion of the historical contexts out of which ISC frameworks emerged
- A set of best practices for doing ISC, including how to consider your positionality and strategies to ensure accessibility in content, medium, and mode

THE WHAT, WHY, AND WHO OF INCLUSIVE SCIENCE COMMUNICATION

Imagine you've been invited to offer general tips about how to communicate a range of data—about the safety of drinking water, air pollution, and housing displacement—to community members in East Palo Alto, a historically marginalized community. You are in the office of a local organization that helps train East Palo Alto community members to advocate for themselves. The room is filled with nearly two dozen people, half of whom

DOI: 10.1201/9781003520887-4

are from East Palo Alto and the other half of whom are university program directors, students, and instructors from your institution. Toward the end of your presentation, the organization's director gets up and walks over to you holding a plastic water bottle, sans label, containing what looks like a clear fluid. She asks, "What do you think is in this bottle?"

You are confused by the question. But you say it is water. It is hot, and you are thirsty. You would drink it without thinking twice. "It looks like water to me," a student agrees confidently. Just then, a community member says, "I'm not drinking that! No way!" Another community member agrees, "No way that's drinkable. I've seen too many bottles like that." Another one joins in: "They say it's good, but it isn't."

You watch, one by one, as each person sees something different in the same bottle. Each person is bringing their past experiences, knowledge, and expertise to the room, which informs how they are interpreting and responding to what is shown to them. At the end of the day, it did not matter what was actually in the bottle or how thirsty anybody was. Only half the room thought the liquid in the bottle was safe to drink, and none of them were from East Palo Alto, whose community members have a long history of their water smelling and tasting bad. According to a needs assessment report, many families were spending more than 40 dollars a month on bottled water instead of groceries (San Francisco Estuary Partnership, 2022). The director holding the bottle said that we all made her point: it didn't matter how many general science communication tips you had diligently prepared to help community leaders translate scientific data. If you are going to be effective, you need to first build trust and credibility by being attentive and responsive to the social, political, and cultural contexts and experiences of the communities engaging with the information.

The lack of trust in science communication has a much longer history that extends far beyond communities in the United States. The exclusion and erasure of the knowledge, lived experiences and expertise of local communities, particularly in the Global South and among Indigenous peoples all over the world, has marked the Western European scientific enterprise for more than 500 years. Elizabeth Rasekoala, President of African Gong, The Pan-African Network for the Popularization of Science and Technology and Science Communication, and Lindy Orthia, a senior lecturer in science communication, note that geologic and astronomical events, as well as mathematical calculations, have long been communicated by Indigenous communities—from Australia to Africa—which have shared environmental and technical information, observations, and applications while passing that knowledge down through generations (*LSE Impact Blog*, 2020). Yet existing science communication histories focus almost exclusively on Western European science. For example, a recent demographic analysis of biology textbooks found that the scientists most commonly featured in textbooks are White men. The study's projections suggest it could take multiple centuries at current rates before we reach inclusive representation (Wood et al., 2020).

It is important to recognize that alongside this history of erasure is a related history of extraction (Bendix et al., 2020). Leading public scientific institutions like the Royal Society in the UK and the French Academy of Sciences, for example, played key roles in shaping scientific discourse to align with colonial interests (*The British Journal for the History of Science*, 1978). The Royal Botanic Gardens and the British Museum exhibitions of "foreign humans" perpetuated the myth of the savage primitive, while research institutions frequently collected data, biological samples, and sacred objects from Indigenous communities without consent. Both Charles Darwin and botanist Sir Joseph Banks traveled on voyages of British conquest, and Darwin's *On the Origin of Species* does not credit any learning from locals whose ancestors had lived in the Americas for nearly 3,000 years (Adame, 2021).

These histories are important for science communicators to learn because they are not relegated to the past. Understanding them helps science communicators attend to how they are still present in our current moment. For example, an analysis of a sample of studies conducted in Africa on a range of infectious diseases found that less than half had an African first or last author (Mbaye et al., 2019). Another article found that two-thirds of high-impact geoscience studies on Africa had no African authors (North et al., 2020), while yet another study found that scientific papers from researchers in the United States, China, and the United Kingdom are more likely to be cited than those on similar subjects from researchers in other countries (Gomez et al., 2022). The lack of diversity is found not only in scientific authorship; it shows up in other ways as well. For example, the percentage of people with disabilities participating in the STEM workforce is significantly low. According to the American Society for Biochemistry and Molecular Biology, as of 2021, only 3% of the STEM workforce identified as disabled. Research also reveals that the *audiences* scientific studies reach are not very diverse. According to numerous studies, traditional science journalism—print, online, radio, television, and non-mediated forms, like public lectures—mostly attracts audiences who are already knowledgeable, interested in science, predominantly White, and wealthy (Humm & Schrögel, 2020).

As Rasekoala and Orthia argue, science communication must "recalibrate mind-sets, and promote diverse recognitions." But how do we ensure these diverse recognitions include more diverse representation of both scientists *and* the audiences that science reaches? ISC offers an important starting point for building the trust that is foundational to achieving this goal.

Inclusive Science Communication (ISC) as Practice and Product

ISC is both a set of practices and a social movement focused on redressing histories of exclusion and erasure and developing more equitable processes for both communicating *and* reimagining how we think about science. ISC prioritizes inclusion, equity, and intersectionality as central parts of science communication, including the goals, design, implementation, and evaluation of science communication efforts. By placing these values and the perspectives of marginalized groups at the center of science communication, ISC welcomes and celebrates diverse ways of knowing, producing, and sharing knowledge (DiCenzo, C., et al., 2021).

ISC can be a central part of public engagement with the sciences, including informal and formal science learning (labs, classrooms, conferences, and public hearings) and science journalism (Canfield et al., 2020). Collaborators affiliated with the Metcalf Institute, a national training center based at the University of Rhode Island that aims to expand news coverage and public conversations about environmental issues, first developed the framework. They offer these bullet points to explain its tenets and goals:

- Acknowledge how historical and present oppressions and inequities affect public perceptions of and participation in STEMM [Science, Technology, Engineering, Mathematics, and Medicine];
- Reflect on how their own biases influence the design and implementation of their work;
- Increase public participation and sense of belonging in STEMM fields;
- Foster collaborations that better serve communities' priorities and enhance the quality and scope of research projects (DiCenzo, C., et al., 2021).

In the following sections, we'll unpack and contextualize these ideas. It is important to note that ISC can be integrated into both the iterative process and the final product that you create (see genre chapters for examples).

TIPS, TECHNIQUES, AND BEST PRACTICES FOR DOING INCLUSIVE SCIENCE COMMUNICATION

Disrupt the Deficit-Model: Understand Your Audience's Beliefs and Knowledge

As we noted earlier, research over the last decade or so has shown that whether and how people pay attention to a given piece of information depends on their prior feelings about and experiences with the topic (Lupia, 2013). But communication processes have historically failed to take this into account. In the era of global development often identified with the post-WWII techno-capitalist, military-industrial complex led by the US, effective science communication was often reduced in the West to a one-way transmission from sender to receiver; those with access to information were charged with making it accessible to those who did not have access. This meant a one-way transmission of communication whose goal was to help people "progress" from an agricultural to an industrial society, from the premodern to the modern, and from the nondemocratic to the democratic. Mass media was the channel through which "modern" ideas were transmitted, and the success of their transmission was measured in economic terms (Schramm, 1964; Lerner, 1958; Rogers, 1962). The strategy of providing the public with information and assuming it would lead to active and sustained change (i.e., modernity)—a strategy that has become known as the "information-deficit" model—has led scientific institutions to launch top-down climate- and energy-related science communication campaigns pursuing a range of goals (education, awareness-raising, behavior change) that have largely fallen short, if not failed altogether (Bauer et al., 2007). The failures of the information-deficit model were caused by multiple factors but, perhaps most importantly, by a lack of awareness of the intended audience's cultural, political, and socioeconomic contexts, including their previously held knowledge, expertise, and beliefs.

If you know what your audience already believes, you can engage with them from a place that centers their own expertise and knowledge first while also being sensitive to previous harms. In cases where science communication is intended to engage with communities that have traditionally been excluded from science-based information or where trust is minimal due to negative past experiences—for example, the historic exploitation of African Americans in medical research (Nuriddin et al., 2020) —having an awareness of and sensitivity to particular audience needs is absolutely essential for gaining audience trust.

The process of obtaining information about an audience and their belief systems depends on the specific case, but science communicators can begin the process by asking the following questions:

1. What does my audience already know about the topic I am going to discuss today? How did they learn it? Through firsthand experience (i.e., mold exposure, lead levels, air pollution)? Through public school education? Through conversations with family and friends?
2. What do the leaders in this community/region/venue/institution think about the topic/issue I'm going to discuss?
3. What are the goals of my science communication? (What do I want audiences to do/think/feel/believe after I communicate with them? How will their prior beliefs impact what they do with the information?)
4. What will this audience have to change/give up after I share this information? In what ways might it contradict or align with their cultural, social, political, and religious beliefs? Can I shape the framing of my communication to be respectful of these beliefs?
5. In what ways can my science communication acknowledge their previous beliefs even if the information I have to share might contradict or complicate them?

While these questions can sometimes be answered only partially (we are always learning), you can begin to answer them by first asking the person or organization who invited you to speak for some of this context. You might also search for basic community demographic information online. Depending on your topic, you can do an online search using keywords from your topic and the name of the town or community with whom you'll be speaking.

Positionality (Know the Affordances and Limitations of Your Own Identities)

As we noted earlier, ISC prioritizes building trust between communicators and audiences, with special care in communities where this trust has been broken. One strategy is to develop an understanding of your own positionality, an understanding that not only builds trust but also makes you a better researcher. Our concept of positionality describes how an individual's perspective is shaped by their social position, including class, gender and sexuality, racial identity, and other determinants of social privilege. The ability to understand and articulate positionality is connected to the difficult task of acknowledging and deconstructing the dominant narratives and personal privileges embodied in our identities (and we all have many identities!), which shape how we understand the world (Pulido & Peña, 1998).

There are many ways to think through how your positionality informs your science communication, from the way you engage an audience (i.e., do you share personal information about yourself in a public talk to show that you have things in common with the audience and can relate to their interests and lived experiences?), to the questions that drive your research. For example, a researcher who lives near an oil refinery might be inclined to ask if corporations that buy carbon offsets—tradable rights or certificates that fund projects that lower greenhouse gas emissions elsewhere to "offset" their own emissions—are perpetuating inequity by allowing corporations to pay elsewhere for the right to continue polluting locally. Alternatively, somebody who does not live in a polluted community might ask whether there are global climate benefits to carbon offset programs.

Some researchers include positionality statements at the beginning of their papers to explain the motivations for their research, amplify its relevance, and clarify how they are connected to it. For example, the following positionality statement was included at the beginning of a student's research proposal focusing on the effects of glyphosate (a widely used pesticide that controls weeds and grasses) use in Northern California public schools:

> I grew up in a family with a severed connection to the land. During the Cultural Revolution in China, my dad was sent away from home to live on a farm. He missed out on high school, and because the opportunity was taken away from him, he felt he had to quell his dreams. Laboring on the land became something to resent. I think that feeling has transformed in his blood, passing into mine as a constant longing for attachment to a place, a relationship with the land, and the desire for good, reciprocal stewardship. My parents' daughter grew up to know English better than their native tongue. Communication was not easy, and

> the way we all came together was through cooking and eating the same Chinese dishes my parents grew up eating. A confluence of forces (being low income, parents overworking, and unhealthy diets) led to my dad having high risk of heart disease and my mom developing Type II diabetes. I didn't realize it then, but later on food would take on another meaning, as something that not only ties me to family and culture, but also as something vital to health. Glyphosate is a part of this picture, not directly to my experience, but as something that has affected a lot of people's lives and operates within the themes of sustenance, bodily harm, and the connection to land.

The extent to which you make your own positionality visible in your written and oral communication will depend on many factors—from the conventions and expectations of your discipline to your own professional preference. However, even if you choose not to make your positionality public, you can still reflect on it to more authentically and inclusively connect with your audience, to focus and refine the questions you are asking in your research, and/or to become more attuned to what you already know and what you need to learn.

Some questions to get you started:

1. What is your own positionality as a researcher and science communicator?
2. How do your various identities intersect with the questions you are asking and/or the rhetorical choices you are making in the design of your materials?
3. What experiences inspired you to ask the questions you do? How might they influence the way you select and engage with your sources and other materials?
4. What can you see that most others cannot because of your positionality? What particular vantage point do you have that is unique from others? Where might your blinders be?

Prioritize Engaging with Local Community Concerns and Expertise and Ensure Equitable Access to Resources

One effective way to connect with your audience as a science communicator is by linking scientific knowledge to areas of immediate concern to them, such as healthcare or economic issues. These linkages can happen when you share your own experience with these issues in the communication of research findings or as part of a partnership with a community in the co-development of a community-engaged research project. For example, Alexis Wilson, a soil scientist, spent more than 3 years working with the Oakland Unified School District (OUSD), conducting soil tests to detect potential contaminants in community gardens in Oakland public schools. She made sure to communicate the results and, where needed, offered remediation guidance to the community in clear and accessible ways (Wilson et al., 2025). Before she began her research project, Alexis met with leaders at the schools to listen and learn about their priorities and to share her own goals for the project. She always made sure her community partners had access to her data and that it was clearly communicated to impacted stakeholders. Ensuring equitable access to scientific knowledge and resources helped Alexis build lasting trust with school leaders.

Alexis writes:

> Data collection and analysis can be technical, so we ensured our partners were informed of the methods we planned to use, approved of the sample collection process, understood what data analyses would be conducted, and what data we would provide. We were also transparent about the limits of our data collection.

As far as effectively communicating the results, Alexis developed several strategies to communicate with the various stakeholders including staff members from each school and district leaders. First, she informed school leadership and the OUSD project team about fieldwork before and after sample collection was completed. To ensure the safety of students and build a connection with on-site staff, she always signed in and out of the school's main office as a visitor and completed a fieldwork observation sheet for each school that included a description of the garden site; when and where sampling took place, with a detailed map of the site; photos; and field notes. All project documents were provided to community partners in a shared online folder so that everyone was kept informed (Wilson et al., 2025).

The previous three sections explored the process of doing ISC via strategies for understanding your own positionality, understanding your audience and their needs, and respectfully engaging with communities while ensuring they have access to your information—whether that is new data or guidance for solving a particular environmental issue. In the following sections, we keep our focus on prioritizing accessibility, with strategies for ensuring benefit sharing in your science communication efforts.

Benefit Sharing Beyond the Initial Communication Output

Benefit sharing in a science communication context refers to thinking through how the people who actively supported you in your research and writing processes might benefit from your science communication efforts. Stakeholders might include community partners, advisors on the project, and others who helped you design and develop the project. Have you discussed with all of your stakeholders the ways they might benefit professionally from the communication of your work together? This might include the potential for co-authorship on papers; co-presenting at conferences where the work is shared out; inviting community partners to guest speak at your institution; or continuing to serve as an advisor on future research projects. The process of ISC doesn't stop with the development of the content but rather must also include meaningful discussion among all stakeholders about how and when the benefits of the research might be equitably communicated and shared.

Make Use of Relevant Communication Channels and "Every Day" Spaces

ISC isn't only about your own experiences and your audience's needs; it is just as important to consider the space in which you are communicating and the medium and modes through which your communication circulates. Local community radio, podcasts, op-eds, documentaries, and targeted social media may be more relevant to and trusted by the audiences you want to reach than more traditional communication sources such as academic journal articles or publications with paywalls.

Inclusivity-inflected genre awareness reflects and reifies many of the genre principles already described in Chapter 1. In order to best think through the genre that would reach audiences that have typically been marginalized, we've adapted a lesson from our colleague Dr. Shay Brawn. Before you finalize the mode of your communication, you can first think through some examples of appropriate genres and discuss why a particular genre may or may not be a good fit for reaching your audience. Some questions for you to consider:

- Who typically communicates using this genre, and what kinds of audiences typically engage with it?

- What purposes does it usually serve for content creators and audiences? (These might be the same, or they might diverge.)
- Where and under what conditions might one expect to encounter this genre?
- Finally, what kinds of subjects and themes are commonly addressed (and not addressed) in this genre?

You'll also want to reflect on the formal and stylistic features of the genre that would be relevant to an audience for whom science communication has not previously been accessible. You can consider the following questions:

- Are there common structures, designs, elements, or moves that will appeal to this audience?
- Are there certain kinds of content that are included or excluded from this genre?
- Are there stylistic patterns (voice, language, appearance) that would be relevant, engaging, and accessible for the people you are trying to reach?

The importance of considering the accessibility and "reachability" of the space in which science is communicated cannot be overstated. This is especially true for science communication events that require physical participation, such as a public talk or conference. Not everybody has the time, resources, or incentive to travel outside of their communities for information they might not perceive as relevant to them. Humm and Schrögel (2020) encourage science communicators to take their work to "everyday spaces."

> In concrete terms this means organizing events in the district center, the village pub, or vocational school instead of the university or research institute–which often are perceived as closed off areas…specifically for science communication, both a fenced research center with guards and admissions controls and a classicist science building (which already seems to signal on the facade that you can't get very far here without being able to have fluent conversations in Latin) for example do not appear welcoming to many people.

Reaching out to public libraries, cafès, and community centers that are already accessible (for example, they have wheelchair ramps and multilingual offerings) ensures that the space will be familiar and comfortable to the audience you are trying to reach and helps to build trust between you and the community.

PRACTICAL TIPS AND TECHNIQUES FOR INTEGRATING ISC INTO PUBLIC PRESENTATIONS

We'll describe preparing presentations in greater depth in Chapter 9, *Scientists Take the Stage*. But there are resources specifically designed to help you think about inclusivity and trust when giving a presentation. A number of organizations have published thoughtful and thorough accessibility checklists to support science communicators who want to be as inclusive and equitable as possible during their public presentations (and you can find more resources along these lines in the "Further Resources" section of this chapter and Chapter 9 as well). These checklists encompass everything from the visuals on slides to inclusive and appropriate language usage. This section attempts to summarize and synthesize the most salient points. When you create presentation materials, you should consider conducting a peer review with others who can give you feedback.

Visual Representation

- Are people of color, people with disabilities, and others who have been visually underrepresented visible in your presentation?
- Will all communities referenced in your communication see themselves represented in your presentation?
- Did you use culturally relevant examples and case studies that resonate with the experiences and contexts of your audience?
- Does your communication—both oral and visual—show people who have been marginalized as experts and leaders? Or are they only visible as receivers of information and resources?
- In what other ways have you incorporated asset-based elements into your presentation that respect and value the ideas, beliefs, and experiences of your audience?

Do Your Visuals Pass a Color-Blindness Test? Two Tips

- Use high-contrast color combinations like blue and yellow and avoid red-green pairings.
- When presenting, do not refer to elements in your slides by color.

Language Usage

- Where possible, did you use person-first language? Instead of saying "handicapped," for example, you might say "person with a disability."
- Did you avoid language that may be unintentionally "othering," and instead choose language that invites belonging? The Sierra Club offers a powerful example: "Instead of 'we stand with LGBTQI people,' you can say 'We support civil and human rights and human dignity for our LGBTQI staff, members, and supporters'" (Sierra Club Equity Language).
- Is the language sufficiently accessible to be understood by your audience(s)? Have you explained or eliminated technical jargon?
- If there are people in your audience for whom English is not their first language, can your communication be translated into their language?
- Are there closed captions and/or a sign language interpreter?
- Have you used alt text on your slides? (Alt text is descriptive text that appears in place of an image or is read aloud by screen readers for users with visual impairments.)
- Do you have room to make the text larger?

CONCLUSION

ISC is a practice and a movement that aims to redress historical and ongoing exclusions and erasures in science communication by prioritizing and valuing diverse ways of knowing, including Indigenous and community-based knowledge, and diverse ways of communicating. As an inclusive science communicator, you can work to develop more equitable spaces for reimagining how and who is communicating "science." It is important to note that ISC is an ongoing, reflexive, and evolving process. Although long-term evaluation and assessment of ISC efforts are limited, recent studies on the impact of ISC trainings and workshops on students have shown that they have been effective in increasing a sense of self-efficacy and belonging among science communication students, particularly those from marginalized communities (Worthington et al., 2023). Many of the strategies discussed here—understanding your audience's beliefs and expertise, thinking through how your own positionality will inform audience perceptions of you, and learning the histories of erasure and extraction that have informed the pursuit of scientific knowledge and its potential impacts on members of your audience—can occur well *before* you begin communicating. Indeed, these are more than strategies. Taken together, they are a collection of intentional practices that work to build the trust that is essential for communicating science effectively.

TAKEAWAYS

ISC is both a set of practices and a social movement focused on redressing histories of exclusion and erasure and developing more equitable processes for both communicating *and* reimagining how we think about science.

A foundation of trust is essential: ISC prioritizes developing trust between the science communicator and their audiences. It starts by understanding what the audience already believes, how they came to those beliefs, and how they contextualize new information.

Disrupt the deficit model: The idea that people lack knowledge and just need more information is deeply flawed and contributes to reinforcing the hierarchies that ISC is trying to disrupt. Learning your audience's familiarity with and previous knowledge of your topic will help you communicate with them more effectively.

Positionality matters: Understand how your identity, background, and social location as a science communicator shape how you communicate and how others perceive you. Knowing the affordances and limitations of your positionality will help you to connect more authentically with your audience.

Prioritize community concerns and accessibility: ISC is most effective when it is relevant to the issues, concerns, and priorities of the people with whom you are communicating. In the context of research projects, it is essential to share data clearly, consistently, and transparently.

Effective ISC is more than just an output—the benefits can be shared: Sharing benefits, such as co-authorship and speaking opportunities, helps keep the science communication effort collaborative and ensures that community members are not just sources of scientific data but also co-creators and beneficiaries. In this way, ISC aims to redress disparities in authorship, citation, and access.

Use accessible channels and familiar spaces: Traditional formats (e.g., academic journals, institutional talks) can exclude audiences who are not able bodied or who have not traditionally had access to formal spaces. Instead, consider trusted spaces—physical and digital—from community centers to local radio.

Design for visual and linguistic accessibility: Representation matters: include diverse imagery in your presentations, avoid "othering" language, use person-first phrasing, alt-text, captions, and, when relevant and where possible, provide interpreters and/or translated slides.

FURTHER RESOURCES

DiCenzo, C., Menezes, S., Smith, H., Murray-Johnson, K., Azizi, M., & McDuffie, K. (2021) *Inclusive Science Communication Starter Kit.* Metcalf Institute, University of Rhode Island, Kingston, RI. 20 pp.

Bartel, B., & Bohon, W. (2020). *Science Communication for Social Justice [Webinar].* National Association of Geoscience Teachers (NAGT). https://nagt.org/nagt/profdev/webinars/social_justice/index.html

Bauer, M. W., Durant, J., & Gaskell, G. (2007). The role of communication in addressing the climate change problem. *Public Understanding of Science, 16*(4), 389–403.

Cite Black Women Collective, a leading citational justice movement to make scholarly references more fair. Retrieved from www.citeblackwomencollective.org/

Diverse Voices in Science Journalism, The Open Notebook. https://www.theopennotebook.com/diverse-voices-in-science-journalism/

Polk, E., & Diver, S. (2020). Situating the scientist: Creating inclusive science communication through equity framing and environmental justice. *Frontiers in Communication, 5*, 6. https://doi.org/10.3389/fcomm.2020.00006

Rasekoala, E. (Ed.). (2025). *Race and sociocultural inclusion in science communication: innovation, decolonisation, and transformation* (Contemporary Issues in Science Communication). Bristol University Press.

Schwabish, J., & Feng, A. (2021). *Do No Harm Guide: Applying Equity Awareness in Data Visualization.* Urban Institute. https://www.urban.org/research/publication/do-no-harm-guide-applying-equity-awareness-data-visualization

REFERENCES

Guralnick, S.M. (1978). [Review of the book The Pursuit of Nature: Informal Essays on the History of Physiology, edited by Alexandra Oleson & Sanborn C. Brown] *The British Journal for the History of Science, 11*(1), 69–71.

Adame, F. (2021, June 16). Meaningful collaborations can end 'helicopter research'. *Nature.* https://www.nature.com/articles/d41586-021-01795-1

American Society for Biochemistry and Molecular Biology. Maximizing Access Committee. (2024). Accessibility in science. *American Society for Biochemistry and Molecular Biology.* https://www.asbmb.org/membership/accessibility-in-science

Bendix, D., Müller, F., & Ziai, A. (2020). *Beyond the master's tools? Decolonizing knowledge orders, research methods and teaching.* International Relations and Colonial Questions. Rowman & Littlefield Publishers.

Canfield, K. (2021, September 13–14). *Inclusive science communication: A framework to help individuals and institutions.* Association for Independent Research Institutes, Virtual conference.

Canfield, K. N., Menezes, S., Matsuda, S. B., Moore, A., Austin, A. N. M., Dewsbury, B. M., Feliú-Mójer, M. I., McDuffie, K. W.B., Moore, K., Reich, C. A., Smith, H. M., & Taylor, C. (2020). Science communication demands a critical approach that centers inclusion, equity, and intersectionality. *Frontiers in Communication, 5*, Article 2. https://doi.org/10.3389/fcomm.2020.00002

DiCenzo, C., Menezes, S., Smith, H., Murray-Johnson, K., Azizi, M., & McDuffie, K. (2021). *Inclusive science communication starter kit* (20 pp.). Metcalf Institute, University of Rhode Island.

Gomez, C. J., Herman, A. C., & Parigi, P. (2022). Leading countries in global science increasingly receive more citations than other countries doing similar research. *Nature Human Behaviour, 6*(7), 919–929. https://doi.org/10.1038/s41562-022-01351-5

Humm, C., & Schrögel, P. (2020). Science for all? Practical recommendations on reaching underserved audiences. *Frontiers in Communication, 5*, Article 42. https://www.frontiersin.org/journals/communication/articles/10.3389/fcomm.2020.00042/full

Lerner, D. (1958) *The passing of traditional society: Modernizing the Middle East.* Free Press.

Lupia, A. (2013). Communicating science in politicized environments. *Proceedings of the National Academy of Sciences, 110*(3), 1291–1297. DOI: 10.1073/pnas.1212726110

Mbaye, R., Gebeyehu, R., Hossmann, S., Mbarga, N., Bih-Neh, E., Eteki, L., Thelma, O-A., Oyerinde, A., Kiti, G., Mburu, Y., Haberer, J., Siedner, M., Okeke, I., & Boum, Y. (2019). Who is telling the story? A systematic review of authorship for infectious disease research conducted in Africa, 1980–2016. *BMJ Global Health*, *4*(5), e001855

North, M. A., Hastie, W. W., & Hoyer, L. (2020). Out of Africa: The underrepresen tation of African authors in high-impact geoscience literature. *Earth-Science Reviews, 208*, Article 103262. https://doi.org/10.1016/j.earscirev.2020.103262

Nuriddin, A., Mooney, M., & White, A. (2020). Reckoning with histories of medical racism and violence in the USA. *NIH Public Access*. https://pmc.ncbi.nlm.nih.gov/articles/PMC7529391/

Pulido, L., & Peña, D. (1998). Environmentalism and positionality: The early pesticide campaign of the United Farm Workers' organizing committee, 1965–71. *Race Gender Class*, *6*, 33–50.

Rasekoala, E., & Orthia, L. (2020). Anti-racist science communication starts with recognizing its globally diverse historical footprint. *LSE Impact Blog*. Retrieved from https://blogs.lse.ac.uk/impactofsocialsciences/2020/07/01/anti-racist-science-communication-starts-with-recognising-its-globally-diverse-historical-footprint

Rogers, E. (1962) *Diffusion of innovations*. The Free Press.

San Francisco Estuary Partnership. (2022). *San Francisco Bay Area Disadvantaged Communities and Tribal Involvement Program regional needs assessment*. San Francisco Estuary Partnership. https://www.sfestuary.org/dactip-regional-needs-assessment/

Schramm, W. (1964) *Mass media and national development: The role of information in the developing nations*. UNESCO/Stanford University Press.

Qureshi, S. (2014, August 6). *Exhibit B puts people on display for Edinburgh International Festival*. The Conversation. Retrieved from https://theconversation.com/exhibit-b-puts-people-on-display-for-edinburgh-international-festival-30344

Wilson, A. M., Polk, E., Field, C. B., & Fendorf, S. (2025). Towards environmental justice: A framework and strategic approach for implementing community-based participatory research in the earth and environmental sciences. *Environmental Science & Policy*, *167*, Article 104036. https://doi.org/10.1016/j.envsci.2025.104036

Wood, S., Henning, J. A., Chen, L., McKibben, T, Smith, M. L., Weber, M., Zemenick, A., & Ballen, C. J. (2020). A scientist like me: Demographic analysis of biology textbooks reveals both progress and long-term lags. *Proceedings of the Royal Society B: Biological Sciences*. . https://doi.org/10.1098/rspb.2020.0877

Worthington, L. A., Miller, K., Winkle, J., Canfield, K. N., et al. (2023). Measuring the impact of inclusive science communication training. *Frontiers in Education*, *8*, 1173661. https://doi.org/10.3389/feduc.2023.1173661

PART 2

Genres for Engaging Public Audiences

CHAPTER 4

Self and Science

Using Personal Narratives to Communicate Science[1]

WHAT YOU'LL LEARN IN THIS CHAPTER

- How scientific knowledge and personal narratives can intersect
- A definition of personal narrative, especially in contrast to the academic article, in terms of idea, evidence, and presence
- A process for developing your own personal narrative, from keeping a notebook to integrating scientific research to revising for voice

INTRODUCTION

Personal narratives engage us with the lived experience of another human being, told directly by them in the first person: an "I" controls the sentences and thus the perspective. In the best personal narratives, we are invited to enter into a life, and through it, a world comes into focus. These narratives come in every length, from micro memoirs of just a few sentences, such as "Tiny Love Stories" in *The New York Times*, to monumental two-volume reflections on a life, such as *A Promised Land* by American President Barack Obama (Obama, 2020).

[1] The title for this chapter – "Self and Science" – as well as the idea that personal narratives are an important part of a science communication curriculum come from an intermediate writing course first devised by Dr. Clara Lewis at Stanford University in 2019.

DOI: 10.1201/9781003520887-6

Scientists often choose to render their research process in personal terms to make their methods and findings more accessible to a general audience. A recent example of this is physicist Claudia de Rham's 2024 book-length account of the place of gravity in her personal and professional life, *The Beauty of Falling: A Life in Pursuit of Gravity.* de Rham describes her development as an intellectual and how she came heartbreakingly close to becoming an astronaut, ultimately linking her experience to a new theory of "massive gravity," a breakthrough that extends Einstein's general theory of relativity (de Rham, 2024). de Rham reviews how scientists have understood gravity historically—from Sir Isaac Newton to Stephen Hawking—and explains why we should care about it, using examples from her own life. In other words, she teaches her readers about the fundamentals of physics and some of its most recent advances.

But even if you haven't made a significant contribution to the area of science you're most passionate about, you still might find that rendering your knowledge in personal terms makes it more attractive, usually because you become a character in the story as a learner of the science with whom your reader identifies. Journalist Rebecca Skloot takes this approach in her bestselling 2017 book *The Immortal Life of Henrietta Lacks*, which follows a suite of cells taken without Lacks' permission through several eras of scientific breakthrough. Skloot layers the racist history of medical science, her interviews with Lacks' family, and explanations of the basic and applied microbiology research with her own first-person, frank reflections on her unpredictable process and self-doubt, making for an educational and compelling narrative.

ARTICLE VS. PERSONAL NARRATIVE: THE GENRE OF THE PERSONAL NARRATIVE IN THE CONTEXT OF SCIENTIFIC COMMUNICATION

We've already addressed the rhetorical tool of "story" in Chapter 2 of this volume. Let's think further about how and why a scientist might use a dedicated first-person narrative to reach a general audience. Ecologist Lauren Oakes published a number of scientific articles on her dissertation research at Stanford University on the effects of climate change on trees and people. Together with several other scientists, she published one in 2016 in *Ecology and Society*, "'I know, therefore I adapt?': Complexities of individual adaptation to climate-induced forest dieback in Alaska," which follows the conventional structure of a scientific article: IMRD (Introduction–Method–Results–Discussion). The article is preceded by an abstract, which summarizes the motive, findings, and conclusion, while the introduction situates the hypothesis within a scholarly conversation, with almost every clause followed by a parenthetical citation of one to three academic sources. The claim—that people adapt to the effects of climate change, specifically to tree dieback, through shifts in action and psychology—is then carefully supported not only through a detailed presentation of the findings but also through a discussion that highlights trends within the findings and stays very close to the facts (Oakes et al., 2016). The language is specialized and dense. Noun phrases pile up, such as the one in the subtitle: "climate-induced forest dieback in Alaska." Feeling is absent. These are not critiques of scientific prose. Rather, it's important to acknowledge its limitations and the reasons why it may simultaneously appeal to readers who trust its objectivity and alienate others who find its dry, jargony style hard going.

Because Dr. Oakes wants the public to understand how climate research is conducted and matters, in 2018, she published a book-length award-winning memoir of her dissertation research, *In Search of the Canary Tree: The Story of a Scientist, a Cypress, and a Changing World*. The memoir doesn't replace the articles that came out of her dissertation because those are important for providing empirical contributions to the scientific community. Rather, it supplements them with her personal perspective even as it continues to report on her scientific findings. Oakes reaches a broad audience through story and arresting language. As she says in her opening line, "I came to Alaska looking for hope in a graveyard" (Oakes, 2018). The idea of the book—that scientific hope can reflect and drive community resilience—is not summarized in a main claim at the end of the introduction but must be experienced across the memoir to be fully understood. Instead of being organized according to the conventions of a formal article, the memoir is largely chronological. Instead of anonymized research subjects, Oakes' memoir is full of very specific human voices, all of whom contribute to her evolving research agenda and approach. Rather than appearing in a table of representative quotes like the one we encounter in her article, the voices of Oakes' informants are represented naturalistically through dialogue. Throughout, Oakes reflects explicitly on her setbacks and successes, not just in science but also in her personal life, as she shares the impact that months spent with a small team of researchers in the Alaska wilderness had on her. Readers gain a holistic understanding of the impersonal "we" of the scientific article and of Oakes as an intrepid, creative thinker (see Table 4.1).

TABLE 4.1 Summary Comparing the Research Article with the Personal Narrative

Article	Personal Narrative
Idea summarized in introduction	Idea developed throughout piece
Topically very specific	Free ranging, experimenting with subject, theme, sources of information, and point of view
Stays close to the facts, rarely strays from the point or interrupts the flow of information with personal opinion	Reflective, unhurried, associative, digressive
Follows conventions of a discipline or academic discourse community	Often unconventional and thus formally inventive
A strong convention is IMRD argument structure: Introduction, Methods, Results, Discussion	Will often use chronology as an organizing principle but may experiment with structure
Quotes, if used, refer to language in published sources or to evidence that is coded as qualitative research	Human voices are present, as the narrator in the first person and, potentially, as characters participating in the story through naturalistic dialog
← Continuum of argument from more formal to less formal →	

While a distinction can be drawn between the research article and the personal narrative, the two genres might be better understood as existing on a continuum of argument, from more to less formal. In both works, Oakes uses the first person, though in her research article it is articulated as the plural *we* since she is working with a team of researchers. She also shows that communities are adapting to climate change in both works. In the memoir, however, we gain a stronger sense of the individual people working to protect forests and communities in Alaska; in the research article, these behavior changes are operationalized—formalized into specific attitudes or actions, such as "promoting: integrating dieback knowledge into professional practices"—so that the research might be replicated. Still, in both genres, the writing expresses the shared keen observation skills of the scientist and the science writer.

Scientists at all stages of their careers, including student scientists, may choose to write a personal narrative for varying reasons. Graduate schools, especially medical schools, require personal narratives in admissions portfolios to assess candidates' motives and preparedness. More established scientists also turn to personal narratives, even in specialist journals. For example, in *JAMA: The Journal of the American Medical Association*, medical doctors can publish personal narratives in a long-running column known as "A Piece of My Mind," which is "devoted to telling stories about the joys, challenges, and hidden truths of practicing medicine in the modern era," according to the website. These last two examples show that scientists use personal narratives to communicate with other scientists, but they also use them to reach a broad public through mainstream publications such as *The New Yorker*, where, for example, Professor of Medicine Atul Gawande addresses issues in public health and surgery, often in personal terms. Unlike specialist academic journals, which are usually subscription-only and locked behind a paywall, magazines are widely subscribed to by public libraries or are open access.

Despite the broad reach and appeal of personal narratives, you may be fearful that relying on the first person will undermine your credibility as a scientist. Scientists working in fields like civil engineering or chemistry, just to take two examples, rarely use the first person to make claims, let alone rely on personal narratives, and thus may have farther to travel before being convinced that their science can be served by this approach. However, scholars increasingly argue that the first person is widely accepted even in formal academic writing. Helen Sword's research has shown that the use of the first person is widespread in scholarly publications across disciplines (Sword, 2011). Some disciplines, particularly in the social sciences, already make storytelling part of their formal methods. Moreover, consider the perspective of award-winning science journalist Ed Yong, writing in *The Atlantic* in 2021:

> a scientist's interpretation of the world is influenced by the data she collects, which are influenced by the experiments she designs, which are influenced by the questions she thinks to ask, which are influenced by her identity, her values, her predecessors, and her imagination.

In other words, our perspectives, even when developed through the scientific method and rendered in specialist prose, are partial, situated in a particular moment in time and place, and reflective of what's more broadly known as our positionality, the social contexts that influence our identities and perspectives (see Chapter 3 for more on this idea). A personal essay, written necessarily in the first person, then, is consistent with science communication in general.

If you undertake a personal narrative, you will want to share some of your educational and cultural history to explain your motives and approaches as a scientist and writer. One advantage of this technique is that it can make more systems of knowledge relevant to science and science communication, showing respect for more communities and inviting a broader audience to your work. Robin Wall Kimmerer, a professor of environmental biology, wrote the bestselling book *Braiding Sweetgrass*, which addresses "the teachings of plants" from both Indigenous and Western scientific perspectives (Kimmerer, 2013). A member of the Citizen Potawatomi Nation, Kimmerer uses personal narrative to share with her reader not only how she came to be a biologist but also what Native values bring to the study of biology. A particularly inventive chapter, "Mishkos Knowmagwen," blends the conventional structure of scientific articles, IMRD (which we saw above in Lauren Oakes' research article), with the story of one of her graduate students, Lena, who worked to test native wisdom about grass planting empirically. Lena's dissertation committee actively resists the research question because it emerges from native knowledge rather than prior published research, but its members also come to approve her work when the careful study results in a new understanding of harvesting practices and crop regeneration (Kimmerer, 2013). In other words, drawing on more funds of knowledge produces better science. Moreover, Kimmerer's first-person account is formally inventive; that is, it both draws on and departs from the conventions of the academic article and personal essay genres. For example, in the "References Cited" section, a nod to mandatory academic citation, Kimmerer lists, with all sincerity and good humor, just four sources: "Wiingaash, Buffalo, Lena, the Ancestors." The chapter "Mishkos Knowmagwen" shows that diverse knowledge traditions can and should contribute to science and science communication, and it models the kind of creative risks you can take with a personal narrative, whose form continues to evolve.

CONVENTIONS OF THE PERSONAL ESSAY GENRE

The notion of risk brings us to the genre of the essay, a term with which you are probably familiar. Personal narratives are often cast as "personal *essays*," which are typically 2–10,000 words, significantly shorter than the book-length personal narratives we have been discussing and therefore more easily produced and better suited to science communication that is urgent. But what exactly do we mean by *essay*? The term comes from the French *essai*, Michel de Montaigne's (1533–1592) word for the new kind of writing he was producing almost daily: reflections on his reading combined with his own philosophizing about his experiences, all rendered in the first person. It was new enough that he thought of this writing as *an attempt*, *trial*, or *test*, some of the meanings of *essai* in French. That spirit of experimentation, combined with close observation, makes the personal essay akin to the scientific process.

For Nicole B. Wallack, director of Columbia University's Writing Program, the genre of the essay has three crucial components: *idea*, *evidence*, and *presence* (Wallack, 2017). The *idea* of an essay is more than a claim or thesis, though these are important. The idea cannot be reduced to a single summary sentence or paragraph, as it must also be experienced across the essay to be fully understood. Tim Requarth's "The Final Five Percent," an award-winning 2019 science essay in the journal *Longform*, documents shifts in Requarth's thinking as he learns more about the implications of his brother's traumatic brain injury. A neuroscientist, Requarth wanted to believe in the power of science to make judicial systems fairer. But this faith was challenged when his brother Conway became increasingly unpredictable and criminal as a result of repeated brain injuries. In his essay, Requarth reviews what the scientific community knows about the relationship between brain injury and self-control and awareness, as well as the checkered history of the American legal system's relationship to brain science, ultimately deciding that brain science in sentencing decisions does more harm than good (Requarth, 2019). This idea is not made explicit at the start of the essay; instead, Requarth postpones it so that we can see and feel the wisdom of this conclusion in the full context of a family and a history.

Even in genres designed to communicate to public audiences, ideas—including those built around a narrative—are only persuasive if they are supported by evidence. Our *evidence* responds to our readers' question, "how do you know that?" It can be seen, heard, tasted, felt, or otherwise accepted by everyone as a remembered fact, and it will be rendered in varied ways in the essay. In the personal essay, ideas will always be supported, at least in part, by lived experience, told by the person who experienced it. For example, in her essay "Dying to be Competent," an explanation and indictment of high rates of black infant mortality in the United States, social scientist Tressie McMillan Cottom tells the story of her pregnancy and daughter's stillbirth (Cottom, 2019). Notice how she also supplements her personal experience with evidence that proves her story is part of a larger pattern, choosing to quote an expert and footnoting a report from the U.S. Centers for Disease Control and Prevention. As she writes: "'Black babies in the United States die at just over two times the rate of white babies in the first year of their life,' says Arthur James, an OB-GYN at Wexner

Medical Center at Ohio State University in Columbus. When my daughter died, she and I became statistics" (Cottom, 2019).

Wildlife biologist J. Drew Lanham's "Forever Gone" likewise draws on varied bodies of evidence to explore the extinction of *Conuropsis carolinensis*, the Carolina parakeet. To conjure an absence, Lanham vividly describes the birds, which continue to exist in art and taxidermy, and documents their difficult history with humans (Lanham, 2018). As you read this extended example, consider why Lanham might choose another type of evidence for his essay, one that has roots in literature and philosophy: the imagined scene. Why does Lanham invite his readers to imagine the landscape fully inhabited by the "gone birds," the species that have gone extinct, at the same time that enslaved African Americans may have been seeking their freedom? Put differently, how does evidence support the idea of this passage?

> I can see the flock circling once, twice, and then on the third pass settling into an old field gone fallow. The unplanted plot is a weedy unkempt affair choked with ragweed and cockleburs. ... They are mostly welcomed there as they digest the undesirable weed in a form of service that the farmers—and the enslaved people working in the fields—appreciate. ...
>
> I wonder about the enslaved watchers who worked in the shadows of endless passenger pigeon flocks that passed overhead, or heard the ivory-bills that called from the tall timber, or glanced at the Carolina parakeets that flashed across work-weary eyes. I think about salt marshes modified and maintained by enslaved Senegalese—the spring crop waving thick with Carolina gold rice and sea-island cotton that created some of the richest men in the world. Beyond the monotonous thud slice of a hoe in pluff mud, what would my ancestors have noticed? The day-to-day work of survival required more than brawn and will. From where would hope emerge? Humans have always looked skyward for inspiration, imagining themselves unbound by gravity or the weight of oppression. Flight means freedom. It is not beyond the oppressed to lean hard on natural beauty as an uplifting beam. ...
>
> For the enslaved, there was more they would have noticed about the parakeets than just their beauty. The parakeets lived in tight-knit social groups comprised of relatives. Along with crows and jays, the parakeets would have been the avian intellectuals in

> a landscape where they had to make do with not only what was, but with what was to come. …
>
> I imagine the flock, assaulted and driven out of a ripening orchard of plums and pears. On the edge of day, as katydids call dusk in, the birds retreat to the forests to roost in cavernous hollows. A depleted flock re-enters a shrinking swamp domain—remnants of wild southern places already disappearing along with the parakeets. They are home, finding comfort in the shadows of last light. But daily, it seems, in smaller and smaller numbers and in fewer and fewer acres.
>
> The birds aren't alone in the refuge. There are humans in the shadows, too. Torches lit from knots of fatwood throw long shadows onto the hollow trunks where the parakeets roost. The dark forms are Maroons, self-liberated slaves who once worked the same plantations that the parakeets frequent.

This passage offers a wonderful example of what Philip Lopate calls "contractions and expansions of the self" in the personal essay (Lopate, 1995): Lanham takes us from his office, an individual, professional space, to a regional landscape in order to make us care about the extinction of one species of bird. Its loss is bound up with systems of oppression that devalue life, including human life. In your essay, you will want to balance your discussion of your personal experience with larger themes or issues. If your readers get the feeling that you're there only to explore a single life, or worse, to indulge your ego, they won't trust your presentation of the science or find meaning in it.

All persuasive scientific communication articulates ideas defended with evidence. What arguably makes the personal essay distinctive is the heightened *presence* of the writer. Consider three examples from the Tim Requarth essay about traumatic brain injury that we discussed previously. What do you notice about the word choice, sentence style, and even punctuation of these passages? Here's Requarth reflecting on his relationship with his brother Conway:

> When I was a teenager and Conway was in his 40s, our father's mind fogged over with dementia and Conway became the closest thing I had to a dad. On a family vacation, he showed me how to drive on country backroads before I turned 16 (I hit a mailbox). A few years later, he taught me how to ride a motorcycle in a

> parking lot (this time, a tree). Conway didn't see the point of many social conventions, which as a teenager I found refreshing.

And the description of one of Conway's devastating injuries:

> The real threat, the injury warranting a helicopter ride to the closest hospital with a neurosurgeon on call, was a hemorrhage beneath the subarachnoid membrane, a thin sheath of triple-helixed collagen fibers intertwined with blood vessels that protects the brain's private chemical harbor of cerebrospinal fluid from the open waters of the body's blood. The sons of a doctor ourselves, my brother and I had heard stories about neurosurgeons called in at midnight, and those stories didn't have happy endings.

Finally, Requarth quotes, among others, nineteenth-century medical reports on the famed traumatic brain injury of Phineas Gage, a railway worker whose brain was pierced by a rod. The injury didn't kill Gage, but it did dramatically change his personality, a result that has fascinated neuroscientists ever since. For example, according to his doctor, John Martyn Harlow, whom Requarth quotes, Gage had been "a great favorite" with his men and in possession of "a well-balanced mind … a shrewd, smart business man." After the accident, he was "fitful, irreverent, indulging at times in the grossest profanity … impatient of restraint or advice when it conflicts with his desires … at times pertinaciously obstinate, yet capricious and vacillating, devising many plans of future operation, which are no sooner arranged than they are abandoned" (Requarth, 2019).

When we read a strong personal essay exploring the sciences, it's as though we're literally in the writer's presence, listening to them talk with feeling and frankness. Conversational style on the page blends specialist and everyday vocabularies, technical terms with the language of the dinner table or front porch. Presence is also cultivated through the writer's use of quotation—who is being cited and how. Harlow's period-specific language works in a number of ways in Requarth's essay. It provides a firsthand account of the consequences of brain injury (so, more evidence) and adds richness to the narrative's style: more voices mean the reader encounters a greater variety of vocabulary and idiom, which creates the experience of a lively conversation.

Through the presence of the writer in the essay, we come to understand their motives and point of view, increasing the impact of the essay's ideas. In Requarth's essay, two selves are strongly present: the science writer, reflecting on what he's learning as he synthesizes his feelings with the medical and legal content, and his brother, a richly drawn character in the story whose struggles with traumatic brain injury are presented holistically and with sensitivity and whose voice speaks to us directly through the liberal use of quotation. While deeply personal—a tale of a single family told by a family member—the message of the essay is not undercut; indeed, we are motivated to engage with the technical content because we come to care about the people who are impacted by it.

CONCLUSION

As we have seen, scientists use personal narratives to contemplate, educate, and advocate. Although the subjectivity of personal narratives might seem antithetical to the objectivity of science, they can still be a very effective tool for communicating science precisely because they are personal, deeply relevant to at least one human and thus likely to many others. When constructed with care—without too much ego or sensational language—personal narratives offer a general audience insight into how science is conducted and often its ramifications for humanity. Moreover, in the age of chatbots, the personal narrative offers an appealing antidote to the generic, disembodied, simulated voice of artificial intelligence. Instead, personal narratives give readers special access to a scientist or an enthusiast learning and working toward new knowledge, an exciting endeavor that keeps the attention of other inquiring minds.

A PROCESS FOR COMMUNICATING SCIENCE THROUGH PERSONAL NARRATIVE

Because it taps so explicitly into the curiosity and passion of a particular self, the personal essay often begins with frequent, informal writing in a journal or notebook. This approach can help you find something to say in all genres but is especially useful for composing a personal essay.

You might think about keeping a journal or notebook as a chance to cultivate your skills of observation, as a space where the scientist and the writer in you find common ground. There is impressive precedent for this. Leonardo da Vinci (1452–1519), artist, inventor, and proto-scientist, kept voluminous notebooks that helped him make original connections among his wide-ranging interests. He would draw things he saw, make note of intriguing quotations from his reading, and jot down ideas as they came to him, often returning to prior pages as his thinking about an object or person evolved. You can check out three of his complete notebooks on the website of the Victoria and Albert Museum (link in Further Resources below). His notebooks were less a diary of emotion and more a record of his observations, a looking out. We are suggesting that you combine the two in your notebook and take pleasure in testing connections. Drawing on his extensive research on da Vinci's process, biographer Walter Isaacson recommends, among many other important lessons, that to develop your curiosity you: 1. start with details; 2. observe thoroughly; and 3. practice accurate reflection (Isaacson, 2017). However, don't feel that you need to write backward, as da Vinci did!

Your notebook can be analog or digital. Some writers like to combine the two, uploading photos of their paper journals to a digital site. When you write in your notebook, write in complete sentences and work to link your ideas, but don't censor yourself. The point is to *write to discover*. You can come back and plumb your notebook for language and ideas that might be useful to your personal narrative.

In brief, we suggest that you first collect material in a notebook, select from it, and then reflect on it as a way of stitching together your memories, experiences, and the science. Note that this process may not be perfectly linear. After reflection, you may find yourself turning back to scholarly databases to collect more scientific information or to your memories for more experiences to help you illustrate the science you are most excited about.

Notebook Prompts

While we've numbered these prompts to suggest an order, you may choose to jump ahead or return to an activity you skipped. Note, though, that the earlier prompts are generally about collecting content, which is where most science writers need to start.

1. Do you have a favorite scientist? Consider starting this project by closely comparing two published pieces by this person: one, a research article published in a journal for other scientists, and the second, a personal narrative, whether oral (in an interview or on a podcast) or written and published in a general-interest publication. Ideally, the two pieces would address the same area of scientific knowledge or inquiry. Begin by making a list of what the two pieces have in common and how they differ in terms of content (what they say) and form (how they say it). What creative risks does your favorite scientist take? Then step back and reflect: Why do you think the scientist

elected to use a personal narrative? Why, in your view, does it reach a general audience? What are the limitations of the personal narrative?

2. List some possible essay topics without thinking about them too critically; the point is to generate 5–6 possibilities in each category. You can articulate the connections later. (Why separate science from self, you may be wondering. To take one example, a student wanted to write an essay on the neuroscience of self-control, his passion, and he found that his experiences as a religious and ethnic minority in rural America, aspects of the self that we will often turn to first as we explore our identities, were not as relevant to the science as his love of rock climbing was. Ultimately, he used his experience as a learner and a rock climber to help his reader understand how our brains both support and undermine our feelings of control.)

   SCIENCE: Some areas of scientific knowledge or debate that you might like to write an essay about

   SELF: Hobbies or experiences that define *you* in part OR cultural traditions that matter to you and why (OR both)

3. Looking at your list of potential scientific topics, highlight those on which you have a strong perspective; by *perspective*, we mean an attitude or opinion about the science that not everyone shares. For each one, note a few of the reasons you feel strongly. Then pick one or two scientific topics that you are most excited about and write at least two paragraphs describing where or how these feelings emerged.

   - Try to make the origins of the feelings very specific and concrete. Where were you? Were you reading something, talking to someone, or watching an event unfold? What happened? What was being said? How did you respond then? Now?

   - Reflect on the origins of your perspective. What are the worlds and contexts surrounding your perspective? What communities do you belong to that inform your perspective? What values, practices, or experiences inform your response?

   - As you respond to the first two bullets, you might also reflect on why this topic should matter to a public audience.

If you're having trouble writing out your thoughts, you might talk them through with a friend or interested listener. The partner should take notes and ask clarifying questions. What do they need to know as a representative of a curious, general audience?

4. Describe possible bodies of evidence for your essay. Think broadly, as you can include many types of evidence in an essay, as we've seen. But also think about what your audience might find persuasive. In your journal, brainstorm:

    - *(Sub)topics.* What will your reader need to understand to make sense of the science you want to explore in your essay? (For example, if you're writing about genetic testing of embryos, you may need to distinguish between germline and therapeutic genetic testing.)
    - Research studies that might support your claim with data
    - Diverse points of view or voices—what evidence do they point to as salient to your topic?
    - Small stories, often from history, that help you show continuity or change across time and place
    - Examples
    - Genres (think broadly: e.g., videos, op-eds, social media posts)

5. Once you've identified an area of scientific knowledge on which you have a perspective, annotate a number of relevant research studies. Each annotation should include some of the most important bibliographic information (i.e., the journal and the date of publication) and be 4–10 sentences long.

    - What are the questions, methods, findings, and conclusions of each study? Why do you trust these studies?
    - After annotating the individual studies, step back and consider how they relate to each other. What is the research story? How does one study respond to the previous one? Is there agreement? Why or why not?

6. Now that you've collected a lot of content on your scientific topic and developed a more robust understanding of it, try articulating a stance on the science in other genres, such as a social media post,

photographs, or word art. What does this help you see about your perspective on the science? For example, does it say something about how you hope to change your audience's thinking or behavior? Does it acknowledge a counterargument? Or help you understand something about your audience? Could you fold it into the essay?

7. At this point, you may have a pretty clear sense of what you want your audience to take away from your essay, but if the central *idea* is not yet clear to you, you might try writing in your notebook in response to each of these questions, which we take from what's known as Stasis Theory in rhetoric, and reflect on where there is the most controversy and where you might contribute (note that if you settle on answering any question beyond fact in this list, you will need to address the prior questions):

   - *Fact*: Does the issue exist?
   - *Definition*: What is the problem?
   - *Quality*: Is it helping or harming?
   - *Proposal*: What should we do?

8. To develop your *voice* in your personal narrative, a key component of your *presence* in the narrative, you will want to reflect on your own individual relationship to language—and to the many Englishes (and other languages) we speak and hear every day. In your journal, write for a few minutes in response to each of these questions:

   - How do you speak in private, with your friends and family?
   - What does your internal monologue sound like?
   - How do you speak at your university, at your job, or in "professional" situations?
   - What do you sound like on social media, if you use it?

   Then, what do you notice about how these voices compare and contrast? How and why might you bring some of these "voices" into your personal essay?

9. You may have been working for a couple of weeks or even months, collecting material and reflecting on it in your notebook. You've perhaps got an idea, you've likely got a lot of evidence, and you're working on presence. The last challenge is organization: what to put where and why in your essay. This can be tricky because essays can have many organizational schemes. They are often organic and relatively unstructured (see, for example, the poet Mary Oliver's (2019) essay "Upstream," which reflects on the importance of nature to her creative identity). Or they can be structured according to the logic of other genres, such as the Kimmerer chapter we discussed above.

   You might start by opening a new document to create a rough outline for your essay. Don't worry about enumerating points. Rather, look back at your notebook entries and consider which concepts, insights, data, quotes from other sources, or small stories you might want to bring into your personal narrative. Select the most compelling material and try roughly grouping it by subtopic. Then step back and reflect on the best order.

   Most personal essays have a story structure to them, the "layer cake" approach we discuss in our storytelling section in Chapter 2, with layers of "showing"—following a character who's doing something in a specific place—alternating with layers of "telling"—explaining why the previous scene matters. Science can be communicated in both modes, either in its making as a "show" or as part of the explanation in the "tell."

   Most personal essays also unfold chronologically, and your personal narrative will likely be organized in part if not wholly, chronologically. In other words, you will detail events from beginning to end, from an initial conflict or problem to a final resolution. But you might also combine this approach to structure with others, especially those that follow you as a character who is learning or changing in response to an encounter with science. (There are other possibilities for story structure explored in Chapter 2 of this book.)

   In your notebook, reflect on the story arcs of your favorite personal essays that communicate science. Consider "reverse outlining" an essay, creating a skeleton in notes of its major points and evidence. Then reflect on where the writer "shows" and where they "tell." Where do they integrate science, and how? Finally, explore in your journal what might make a compelling story arc for your essay before turning to your outline to play with various organizational schemes.

10. Bringing it all together in a very first draft of your personal narrative: try recounting a story out loud to a partner.

    - You are the character; the plot relates to the topic in the sciences that you want to explore/communicate (some dimension of it); the setting is where/how it unfolded. Try to detail an "inciting incident"—when the story began—specifically how you came to a question or concern, nailed in time and place. Then show how you think you might resolve it.
    - Take notes for 5–10 minutes before you start talking.
    - Consider recording your first draft and using voice-to-text software to create a first rough draft of your essay.

    Your listeners should take notes while listening. When did they feel most connected to the story? Why? What did they want to hear more about?

11. Consider developing and revising your essay using the following guidelines, written in collaboration with our students in class.

    A successful personal essay that also communicates science will:

    - Develop a compelling idea throughout and support it with varied sources of evidence
    - Reflect in meaningful ways on the writer's positionality in space, time, and/or culture
    - Take some formal risks in organization, style, or content
    - Communicate science accurately while engaging a general audience
    - Transition carefully and strategically from one idea to the next

    You might also write your own "rubric": given your reading and learning style, what are the components of an impactful personal narrative? Try to list these as precisely as you can and then read your draft critically through that lens. Which elements need strengthening and what can you imitate from your self-chosen "model of greatness"?

12. Read your draft out loud. Does it sound "conversational"? If not, see if you can bring more voice to the page by blending every day and specialist vocabularies, shortening your sentences, and making your tone more frank.

TAKEAWAYS

- Personal narratives offer an exciting opportunity to explore the relationship between self and science.
- Scientists and science writers turn to personal essays and memoirs to contemplate, educate, and advocate.
- You can collect your own experiences and observations about science in a notebook before deciding what might be essay-worthy.
- A strong first-person voice and story structure are likely to keep your reader engaged.

FURTHER RESOURCES

Hart, J. (2021). *The complete guide to writing narrative nonfiction* (2nd ed.). University of Chicago Press.

Jorgensen-Earp, C. R., & Jorgensen, D. D. (2002). "Miracle from mouldy cheese": Chronological versus thematic self-narratives in the discovery of penicillin. *Quarterly Journal of Speech*, *88*(1), 69–90. https://doi.org/10.1080/00335630209384360

Another example from the history of science that shows personal narrative can play an important role in making scientific discovery memorable.

Lopate, P. (1995). *The art of the personal essay: From the classical era to the present.* Anchor Books.

Victoria and Albert Museum. Leonardo daVinci's Notebooks. https://www.vam.ac.uk/articles/leonardo-da-vincis notebooks?srsltid=AfmBOooJfQfSF199KLRJFbRm-1dDI2cE2E5_06xjwVw72PdgCq3WGKW3

Online visual resource that links you to the notebooks known as the Codex Forster I-III. Here you can be inspired by daVinci's notebook keeping practices in vivid detail.

REFERENCES

Cottom, T. M. (2019). Dying to be competent. In *Thick and other essays* (pp. 73–98). The New Press.

de Rham, C. (2024). *The beauty of falling: A life in pursuit of gravity*. Princeton University Press.

Isaacson, W. (2017). *Leonardo da Vinci*. Simon & Schuster.

Kimmerer, R. W. (2013). *Braiding sweetgrass: Indigenous wisdom, scientific knowledge, and the teachings of plants*. Milkweed Editions.

Lanham, J. D. (2018, February 21). Forever gone: How bird lives and black lives intertwine in the long shadow of history. *Orion*. https://orionmagazine.org/article/forever-gone/

Lopate, P. (1995). *The art of the personal essay: From the classical era to the present*. Anchor Books.

Oakes, L. (2018). *In search of the canary tree: The story of a scientist, a cypress, and a changing world*. Basic Books.

Oakes, L. E., Ardoin, N. M., & Lambin, E. F. (2016). "I know, therefore I adapt?" Complexities of individual adaptation to climate-induced forest dieback in Alaska. *Ecology and Society*, *21*(2). https://www.jstor.org/stable/26270389

Obama, B. (2020). *A promised land* (Vol. 1). Crown.

Oliver, M. (2019). Upstream. In *Upstream: Selected essays* (pp. 3–8). Penguin Books.

Requarth, T. (2019, October). The final five percent. *Longreads*. https://longreads.com/2019/10/22/the-final-five-percent/

Skloot, R. (2017). *The immortal life of Henrietta Lacks*. Broadway Books.

Sword, H. (2011). *Stylish academic writing*. Harvard University Press.

Wallack, N. B. (2017). *Crafting presence: The American essay and the future of writing studies*. University Press of Colorado.

Yong, E. (2021, October). What even counts as science writing any more? *The Atlantic*. https://www.theatlantic.com/science/archive/2021/10/how-pandemic-changed-science-writing/620271/

CHAPTER 5

Designing Interactive Science Exhibits

WHAT YOU'LL LEARN IN THIS CHAPTER

- Learn about the different kinds of interactive science exhibits, from large-scale, expensive exhibits to small, portable exhiblets that rely on free or reused materials.
- Understand both the opportunities and the challenges of designing effective exhibits that engage visiting audiences in learning about science.
- Follow a process for designing an exhibit, from coming up with a big idea to writing accessible, engaging exhibit labels.
- Discover how exhibit design activities can help you think creatively about structure and audience engagement in other genres of science communication.
- Discover further resources for science communicators interested in exploring exhibit design.

DOI: 10.1201/9781003520887-7

INTRODUCTION

> It all started when I was seven. And I went on a family trip to Chicago, and we visited Field and there was basically a sandbox filled with bones, and you could dig up your bone. And that was the first time I learned what paleontology even was, and I loved it.
>
> —*Paleontologist Lynnea Jackson*

When scientists speak about how they developed their initial interests in science, there is often a childhood experience at a science museum, like Jackson's visit to the Field Museum, a field trip, or a family excursion to a planetarium or nature center that sparked their curiosity or sense of wonder about the natural world. When we ask science communication students in our classes what drew them both to science and to an interest in communicating science across audiences, vivid memories of trips to museums, aquariums, and discovery centers often surface in similar ways. Students describe childhood memories of feeling a starfish at an aquarium touch tank, watching stars mapped into constellations on the ceiling of a planetarium, or understanding a scientific concept like reflection and refraction—the bending of light as it moves through different media—when they had a chance to play with lenses and light beams in a carefully designed, interactive exhibit.

These stories and memories speak to the power of science museums and exhibits to communicate science to the public in ways that extend far beyond the initial visit, prompting further interests and questions. Or, as researchers at the Exploratorium museum in San Francisco put it, "Science museums, when done right, have the potential to positively shape visitors' affective attitudes, understanding, and motivation toward science" (Chien et al., 2025). But how do we create the kinds of engaging science exhibits that will inspire wonder and lasting learning for visitors? How can we design exhibits that are truly interactive and multisensory while also ensuring that interactive features don't overwhelm or distract from our visiting audience's learning? And what can learning about exhibit design teach us about communicating science in other genres and contexts?

GENRE IN CONTEXT: HISTORY AND EXAMPLES

What does it mean for an exhibit to be interactive? In her foundational guide Planning for People in Museum Exhibitions, curator and exhibition director Kathleen McLean describes interactive exhibits as "those in which visitors can conduct activities, gather evidence, select options, form conclusions, test skills, provide input, and actually alter a situation based on input" (McLean, 1993).

While the first mention of a science exhibit might conjure an ambitious, multi-room exhibition hall at a major museum like the Museum of Natural History in New York or Chicago's Field Museum, interactive science exhibits range from expensive exhibits that take years to develop to small, temporary "exhiblets" set up to engage a more specific community in learning for a few minutes. To broaden our sense of what is possible with interactive science exhibits, below are descriptions and examples of interactive exhibits from throughout the San Francisco Bay Area, along with the diverse audiences they engage and the resources they require.

Science Museum Exhibitions

What many of us think of first when we think of science museums and exhibits are ambitious hands-on exhibits within a large science museum or museum of science and industry. Key features of these exhibits include:

- Organized around a theme.
- Resource-intensive.
- Years to develop with a multidisciplinary team.
- Designed to reach thousands of visitors a year across demographic categories and ages.

Example: The iconic "Seeing" exhibition at the Exploratorium museum in San Francisco is an example of this kind of exhibition. Organized around the theme of visual perception, this exhibit features 25 individual exhibits through which visitors can explore color, light, perception, illusion, and camouflage. Younger visitors can jump in front of a screen in the "Shadow Box," leaving still images of their shadows mid-jump on a screen, or interact with a giant curved mirror without needing to read any instructions. Elsewhere in the exhibit, an older visitor might carefully read an entire exhibit label to take on the challenge of splitting light and bringing it back together in a single beam, experiencing and learning about refraction, color absorption, and light gathering (*Seeing | Exploratorium*, n.d.) (Figures 5.1 and 5.2).

FIGURE 5.1 Interactive exhibits on visual perception and the behavior and physics of light in the Seeing gallery at the Exploratorium. (Reprinted with permission from Megan Formato.)

FIGURE 5.2 The Giant Mirror exhibit helps visiting audiences of all ages explore how light bends when bouncing off curved surfaces. (Reprinted with permission from Megan Formato.)

Smaller Exhibits or Displays

As part of a nature center, children's discovery museum, or library these displays are often:

- Centered narrowly on a theme or a place.
- Can have a more focused audience; for example, just children or residents of a certain area who frequent a park or library.

Example: The Palo Alto Children's Museum features interactive exhibits that engage young children in learning about the natural world. Situated between a playground and a children's library and theater, the museum includes exhibits like a sand area where children can dig for replicas of dinosaur fossils and wind tunnels where they can test paper airplanes in flight. A scale model of the solar system begins outside the museum and extends through a local park to the library, communicating the vast empty space between planets in a way that is hard to visualize from a solar system poster, drawing, or exhibit confined to an indoor space. The simple outdoor exhibit includes short, provocative facts about the planets that spark further questions and inquiry among the diverse populations and ages that frequent the playground and library. On any given day, you can see caregivers reading facts and questions about the planets to children and children running from planet to planet, marveling at how far apart the planets are.

TRAVELING EXHIBITS, ONLINE EXHIBITS, MICRO MUSEUMS, AND STANDALONE EXHIBITS

Community Science Workshops and Traveling Exhibits. Community Science Workshops focus on community and foster experiences in which learners work together and collaborate with staff to experiment with, interact with, and modify exhibits (Nee, 2024). Key features include:

- Providing hands-on, accessible STEM learning through free or low-cost activities for youth and families.
- Promoting equity and inclusion in science education by serving underrepresented and underserved communities.
- Developing community connections through mentorship, collaboration with local organizations and schools, and open workshop spaces.

Example: Mission Science Workshop's Whale on Wheels

Mission Science Workshop has three community workshop sites for tinkering and exploration in San Francisco. It also offers traveling exhibits, including its Whale on Wheels program, which brings a real 40-foot juvenile gray whale skeleton, "Gracie," to schools and community partners. After educators guide a discussion related to the whale, students put the whale skeleton together like a giant puzzle, learning about anatomy and evolution. Students learn comparative anatomy, including the similarities between a whale's flipper and a human hand, as well as about different types of whales, whale life cycles, and the impact of factors like overfishing and pollution on whales. Bilingual (Spanish/English) follow-up resources and materials are shared with teachers so that they can extend the learning beyond the visit itself (Whale on Wheels | Mission Science, n.d.).

Exhiblets

In 2024, the Exploratorium teamed up with the Global Community Science Workshop Network to figure out how to bring the kind of learning that happens in an interactive science museum into other community spaces. They use the term "exhiblet" to describe museum-style exhibits that "can be built on a shoestring budget using ordinary tools" and repurposing everyday stuff, such as a hairdryer or trash can (Exploratorium, 2024). Key features:

- Small, low-cost, approachable interactive exhibits made outside of a museum with common, low-cost materials.
- Inspired by the Community Science Workshop approach to building community through open-ended tinkering.

Example: Bernoulli's Hair Dryer

In this exhiblet, developed by Free Science Workshop in partnership with the Exploratorium, a hair dryer affixed to a block of wood can be used to float a balloon or a ping-pong ball, demonstrating Bernoulli's principle in action: the balloon floats because the faster moving air from the hair dryer creates a low-pressure zone in which the balloon rests, while the slower-moving air around it is at a higher pressure, holding the balloon in place. The exhiblet is portable, made of common, inexpensive materials, and easy and accessible to operate (Exhiblets – Free Science Workshop, n.d.).

As you begin to think about the science you want to communicate and the audiences you want to reach, consider the full range of exhibits above. Consider where the visitors you care about reaching will be able to access your exhibit. Is a park, a school, a community center, or a library a possibility? Does your exhibit need to be portable? Designing an interactive science exhibit involves reflecting on the budget, infrastructure, space, and collaborations you may need.

ACCESS AND INCLUSION AS PRIORITIES IN SCIENCE EXHIBITS

Given the mission of science exhibits of all kinds to communicate science in compelling ways to diverse audiences and to offer visitors opportunities to co-construct knowledge through interactive exhibits, questions of accessibility and inclusion are especially important to prioritize. During the second half of the twentieth century, museums moved away from a behaviorist approach—where authoritative knowledge was meant to provoke specific responses and learning from visitors—to constructivist approaches—where exhibits anticipate that visitors will bring their own knowledge and co-construct learning through an active experience. A behaviorist approach to an exhibit about the reflection and refraction of light might involve a diagram or a pre-set physical arrangement of a light beam and a lens, accompanied by a label that explains the physics. A more interactive approach, in which visitors co-construct this learning, could be something like the Light Island exhibit at the Exploratorium, where visitors manipulate beams of light using an assortment of lenses and mirrors. Notice how open-ended challenges and questions invite visitors to experience how light bends and bounces in order to arrive at their own understanding (Figures 5.3 and 5.4).

With this shift from behaviorist to interactive approaches, designing exhibits that value more forms of prior knowledge and ways of understanding the world has become crucial. Unfortunately, as science communication scholar and recent Exploratorium fellow Elizabeth Rasekoala points out, many science museums continue to fail to incorporate cross-cultural examples of science and instead focus on Western science, ignoring their responsibility to share examples from other knowledge systems (Rasekoala & Orthia, 2020). Not only does this exclusion communicate an inaccurate and lopsided view of scientific knowledge and forms of inquiry, but it also creates obstacles to learning for visitors who don't see their own identities and cultures reflected in the version of science that the exhibit projects.

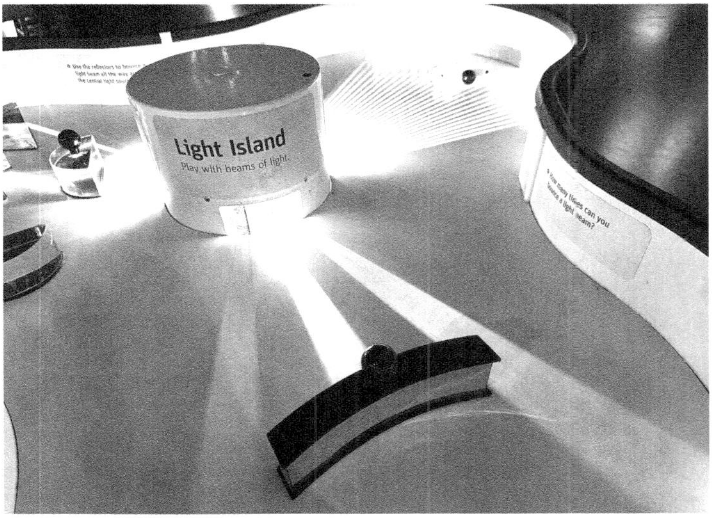

FIGURE 5.3 The Light Island exhibit at the Exploratorium is an example of an interactive exhibit where visitors construct their own knowledge about the behavior of light. (Reprinted with permission from Megan Formato.)

FIGURE 5.4 Open-ended prompts and questions displayed around the exhibit encourage visitors to learn about the bending, bouncing, and reflection of light. (Reprinted with permission from Megan Formato.)

FIGURE 5.5 Día de los Muertos (Day of the Dead) ofrendas (altars) specific to Oaxaca Zapotec, Guatemala K'iche' Maya, Yucatec Maya, and Chicano traditions fill the opening gallery of the Exploratorium in October 2025. (Reprinted with permission from Megan Formato.)

For visitors to have the kind of transformative learning experiences exhibit designers hope for, they need exhibits that represent and engage their own perspectives and ideas as part of the learning process.

The Exploratorium, where Elizabeth Rasekoala was a recent museum fellow, offers several instructive examples of how exhibits can center different visitors' identities and cultural practices and help visitors learn more globally inclusive forms of knowledge. On display in October in the opening gallery space is a temporary exhibition for Día de los Muertos (Day of the Dead). The exhibition features wall labels that explain to visitors that many Indigenous cultures throughout Latin America create altars (ofrendas) and decorate them to honor their ancestors during Día de los Muertos. Detailed exhibition labels also make connections between ofrendas and science, pointing to the relationship between Día de los Muertos and astronomy and connecting Día de los Muertos to social scientific understandings of how we grieve (Figures 5.5 and 5.6).

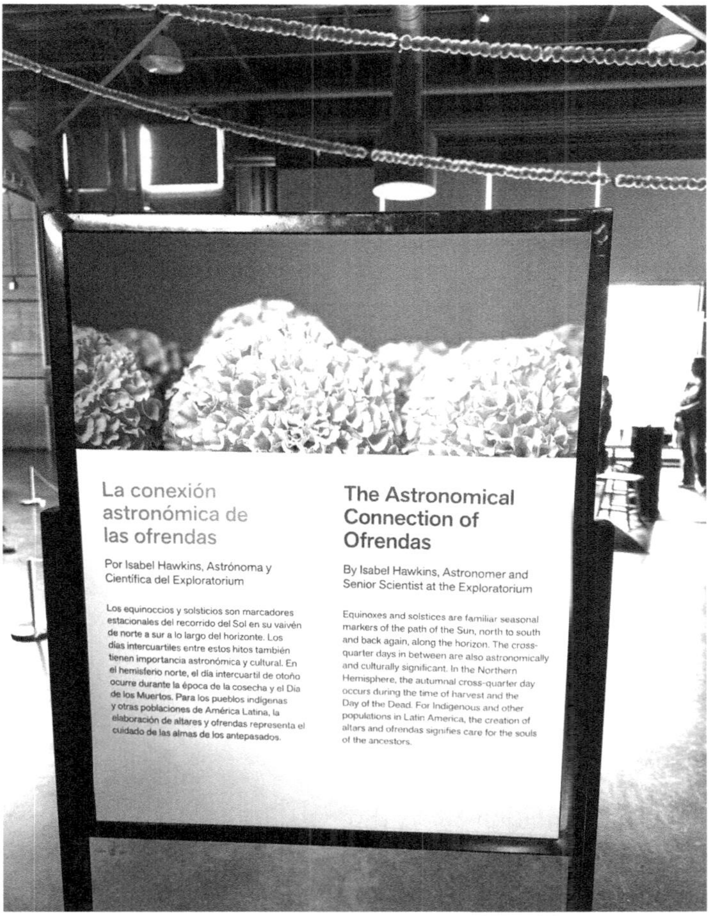

FIGURE 5.6 Prominent labels in Spanish and English describe the connections between ofrendas and scientific knowledge. (Reprinted with permission from Megan Formato.)

Deeper in the museum, a social science–based exhibition includes interactive exhibits that invite visitors to reflect on their own names, emotions, and identities (Figure 5.7).

An Exploratorium exhibit about emotions asks visitors "How do you feel about feelings?" and prompts them to look in a mirror to examine their own emotions and to describe the emotion in shown each photograph on a wall. Notice how the choices of images reflects the diversity of the visiting audience in terms of race, age, and gender (Figure 5.7).

FIGURE 5.7 Inclusive science communication in an exhibition about emotions. (Reprinted with permission from Megan Formato.)

Many of the exhibition labels throughout the Exploratorium are presented in the three languages most commonly read by visitors to the museum, and they also include visual icons to guide visitors who do not read those languages (Figure 5.8). These choices help make the organization of the museum and the purpose of different galleries accessible to more visitors and they help more visitors understand that the museum has been designed with them in mind.

The stakes extend beyond the learning experiences of individual museum visitors. Emerging scholarship in museum studies and science communication argues that if museums don't take up the work of repatriating stolen objects, incorporating cross-cultural examples, and interrupting a Eurocentric gaze, they continue to participate in and further colonial projects. Chapter 3, *Making Science Inclusive and Accessible*, offers more examples of how science communicators can approach inclusive science communication and tackle this vital decolonial work in both theory and practice.

FIGURE 5.8 Multilingual heat and temperature exhibition wall label. (Reprinted with permission from Megan Formato.)

CORE DESIGN PRINCIPLES

The examples above point to the many advantages of science exhibits for communicating science to the public. Science exhibits are accessible to people of different ages and levels of literacy. They are hands-on and engage and encourage different ways of knowing and learning (Allen, 2004). They offer an opportunity to educate within a specific place or community. As open-ended experiences, they can vary the depth of insights and takeaways for different ages of visitors who spend different amounts of time with the exhibit. Well-designed interactive exhibits also often create an opportunity for visitors to experience part of what it is like to be a scientist, including some of the fundamentals of observation, experimentation, and drawing insights from what they see or do.

The many opportunities for engagement and learning that exhibits provide come with challenges that exhibit designers and curators must consider:

Unlike other genres in this book, an exhibit designer cannot predict how a visitor interacts with their exhibit. While an author of a profile or opinion article can assume that a reader might start at the top of the page, an exhibit designer cannot anticipate where a visitor will begin to explore their exhibit, what order they do things in, how long they spend, what they read, and what they don't. There are ways to suggest structure but not control it, so a designer must consider what visitors who spend different amounts of time and engage in varying ways will walk away having learned.

While there is ample evidence that many visitors learn through doing and hands-on engagement, museum studies research also suggests that too much interactivity can overwhelm visitors and contribute to cognitive overload that interferes with learning (e.g., Allen and Gutwill, 2004; Evans, 1995; Hedge, 1995). Exhibit designers can use user-centered design and user testing to ensure that their exhibits are engaging and interactive without overwhelming or confusing visitors (Table 5.1).

TABLE 5.1 Common Interactivity Challenges and Their Fixes

Challenge	Fix
Too Many Options that All Seem Equally Important to a Visitor Too many options for what to do demand the visitors attention at once.	**Make Clear What is Most Important to Interact with and Why** • Use labels and specific questions in a specific order to provide structure and order. • Limit interactive features; ensure that all interactive features have a clear learning goal and purpose related to the exhibit's main idea • Consider user-centered design: knobs are for turning, buttons are for pushing, and handles are for pulling, ensuring that interactive components are intuitive and align with ways that visitors interact with other objects in their day-to-day lives.
Secondary Features Obscure Primary Feature A button needs to be pressed or a container needs to be filled for an exhibit to function, but younger visitors might become absorbed in the button-pressing and bucket-filling and miss out on the learning from the exhibit as a whole.	**Anticipate Initial and Prolonged Engagement** If a visitor just presses the button, what can they see or learn? What can they learn from filling and emptying the bucket? Anticipate both brief initial engagement and prolonged exploration to a visit and ensure that learning can be achieved through both.
Interactive Features are Limited to one Sense or Language	**Anticipate Engagement with Multiple Senses and by Users Reading Multiple Languages**

Let's see how a few of the interactive exhibits at the Palo Alto Children's Museum and Zoo navigate these interactivity challenges through exhibit design:

The Spinning Pattern exhibit invites visitors to pile sand on a spinning circular surface and observe the patterns they can create and the ways in which the rotation of the surface affects those geometric patterns.

While this is an open-ended exhibit that visitors interact with in a myriad of ways, the exhibit label pairs prompts and questions to guide how users engage with the wheel and the sand, inviting them to reflect on what they see or feel. Visitors are invited to "Try drawing a square" and prompted to consider, "What happens?" The wheel spins at a constant speed on its own (Figures 5.9 and 5.10).

Look carefully at the photographs of the exhibit and its labels above. Consider these questions on your own:

- How does the exhibit navigate some of the common interactivity challenges using these principles and strategies?
- How do the label and design make clear what is important to interact with and why?

FIGURE 5.9 Spinning patterns label at the Children's Museum in Palo Alto. (Reprinted with permission from Megan Formato.)

FIGURE 5.10 Instructions for interacting with the exhibit. (Reprinted with permission from Megan Formato.)

- How do the label and design ensure that learning is accessible to different learners and literacy levels?
- How does the exhibit anticipate different lengths of visits? How might a 4-year-old who doesn't read interact with this exhibit during a brief 2-minute visit? What would they do or learn?
- What might an 8-year-old who sits down with it for 10 minutes try and learn?
- How would a grandparent who reads Spanish interact with it together with their toddler grandchild?

A PROCESS FOR CREATING YOUR OWN SCIENCE EXHIBIT

Now that we have considered different kinds of interactive science exhibits, rich examples of exhibits for different audiences and budgets, and core design and inclusive science communication principles, we'll share a process for putting some of these ideas to work in your own exhibit.

Because the process of designing entire science exhibitions is often carried out over years by a multidisciplinary team, the process below does not attempt to guide you through the entirety of that ambitious task.

Instead, through the activities below, we will learn how to develop a guiding big idea for an exhibition, how to approach design and consider our communication goals for specific audiences, how to represent a design to others for feedback and revision, and how to write the different kinds of exhibit labels that guide a visitor's experience.

These steps should be useful for designing an interactive science exhibit, but *many of the activities can be helpful to science communicators working in other genres as well.* Coming up with a big idea might be useful as part of creating a podcast (Chapter 8) or a personal narrative essay (Chapter 4). The sprinter, stroller, studier revision activity, designed to anticipate visitors who spend different amounts of time with an exhibit, could be useful for anticipating a reader who glances, skims, or carefully reads a science op-ed (Chapter 6) or profile (Chapter 7). And sketching an exhibit case of key examples or people can offer helpful new ways of thinking about structure, even if an exhibit isn't your ultimate goal.

Invention and Research

Before you begin, have a preliminary scientific theory, figure, or debate that you know you want to communicate, as well as a sense of the audience(s) you want to engage in learning about it.

Exhibition Reviews/Genre Analysis

Visit Three Science Exhibits That Are of Interest to You

1. *Interactive*: An exhibit that offers you the chance to manipulate objects, touch something, make something, or otherwise respond to the exhibit content.
2. *Descriptive*: An exhibit where you are mostly reading or looking at images, graphics, and objects.
3. *Digital*: An online exhibit.

Spend 30 minutes exploring each exhibit and, if possible, watching others interact with it. Take notes in response to the questions below during or immediately after your visit:

Draw the exhibit and/or list its component parts: Don't worry about the quality of your drawing; this is just a schematic. Label the parts or create a list of the parts of the exhibit. Is there a label? Directions? Photographs? Objects? What is the exhibit made of?

Describe your pathway through the exhibit: What did you do first? Why? What did you do next? You can write this as a paragraph or a list of steps you took. Where did you begin, and why? How did you know you were done exploring the exhibit? Was there a logical end, or did you get bored or run out of time?

Describe others' interactions with the exhibit: If you have the chance to watch other people interacting with the exhibit, what do they read or do? Are they doing anything different from what you did during your visit? Who would this exhibit be most accessible to and why? Who would it be least accessible to?

Describe the exhibit to another visitor: Imagine another visitor is approaching the exhibit and asking you about it just as you wrapped up your visit. Briefly answer their questions about it: What is the exhibit about? What are its big ideas? What do I get to do or see? What will I learn? What do I need to know to understand it?

Reflect on Your Exhibit Reviews

When you consider who each exhibit reaches and the kind of content and learning it offers, do any of these types of exhibits feel like the best fit for either the audience you wish to reach or the scientific ideas you wish to share?

Which did you enjoy the most, and why? Which did you learn the most from, and why?

Coming Up with a Big Idea

> A powerful exhibition idea will clarify, limit, and focus the nature and scope of an exhibition and provide a well-defined goal against which to rate its success.
>
> —*Beverly Serrell, Exhibit Labels: An Interpretive Approach*

Educator and exhibition consultant Beverly Serrell (2015) encourages designers of all exhibits to begin with a big idea. For Serrell, "A big idea is a sentence—a statement of what the exhibition is about. It is one complete, noncompound, active sentence that identifies a subject, an action, and a consequence for the exhibit."

A big idea serves as a guiding thread to give an exhibition coherence, focus, and depth, helping the curatorial team and exhibition designers decide what content to include and leave out. It is meant for internal use rather than public display. Without a big idea, exhibitions often become confusing, overwhelming, and poorly structured, making it difficult for visitors to engage meaningfully. Using a big idea increases the chances that visitors will understand the exhibition's intended message.

Big Idea = Subject + Action + Impact for the Exhibit

(the Exhibit's "So what?")

Examples of Big Ideas:

- What we see can teach us about our underlying anatomy and the behavior of light.
    - This is the big idea that underlies the Exploratorium's Seeing exhibition.
- The presence of feedback systems in organisms and ecosystems helps to maintain homeostasis—equilibrium states in which variables are maintained within healthy parameters.
- A scale model of the solar system highlights the vast emptiness of space and the value of proportional thinking.
    - This is the big idea that underlies the Palo Alto Children's Museum scale model of the solar system.

Examples of unhelpful big ideas. What are they missing?

- This is an exhibition about visual perception.
- Learn everything about the human body.
- Kelp forests are important ecosystems.

Consider your preliminary topic, audience, and the form of exhibit you are considering, and write your own big idea to guide your design process.

Prototype, Empathize, and Ideate

What is a Prototype? Prototyping is an iterative process for creating, testing, and revising exhibits with potential audiences. A prototype is a model that is inexpensive and fast to create and that a user can interact with enough to give you feedback on their experience of your design. A prototype can be a model, a clear sketch, or a digital simulation. Think creatively about how to represent the experience to a user for feedback.

Why Prototype? Prototyping enables you to test different ideas before investing the time and resources needed to build the final working exhibit. It helps you empathize with your audience and understand their experience.

Tips for Prototyping

- You don't have to prototype your entire exhibit. Prioritize interactive elements that you think are most important to your exhibit's success.
- Your prototype doesn't have to have the full functionality of the final exhibit. For example, knobs and buttons may not perform the functions they eventually will.
- A prototype can be a model, a clear sketch, or a digital simulation. Think creatively about how to represent the experience to a user for feedback.
- Before sharing a prototype with someone else for feedback, go through a few rounds yourself. Think about what you know about the visitors you want to reach, imagine their encounter with the prototype, reflect, and iterate.

Museum Visitor Persona Exercise: Sprinter, Stroller, Studier

Exhibit designers know that visitors will spend different amounts of time with an exhibit. While some will spend just a few minutes (a sprinter), others will stay longer and engage deeply with the text and interactive elements you provide (a studier). What do you hope visitors who spend different amounts of time with your exhibit will learn?

Make a timeline for a 5-minute visit to your exhibit and a 10-minute visit. What will the visitor do first, second, and third? Be realistic about timing: how much time does it really take to read instructions or explanations on a label? How long does it take to interact with different elements of the exhibit in different ways?

TABLE 5.2 Anticipate How Different Visitors Experience Your Exhibit

	Sprinter (2 minutes)	**Stroller (5 minutes)**	**Studier (10+ minutes)**
What are your learning objectives for them?			
What do they do and interact with?			
What do they read?			

After you've completed your timeline, more systematically consider how the sprinter, stroller, and studier personas experience your current design. Draw a chart like the one above and complete it for visitors who spend different amounts of time interacting with your exhibit (Table 5.2).

Reflection Questions

When you consider what you set out to teach your different visitors (your learning objectives) and compare that to what you anticipate they would do and read at your exhibit, do you see any gaps? Is there anything they are reading or engaging with that distracts from your goals for what you hoped to communicate? What else do you know about potential visitors to your exhibit that might inform your design? Age range? Literacy levels? Languages?

Consider your answers to these reflection questions as you revise your sketch of your exhibit. Remove elements that are distracting, and incorporate interactive design elements from Table 5.1 to help support your visitors' focus on your learning objectives and help them avoid distraction.

FURTHER RESOURCES FOR EXHIBIT DESIGN

Davidson, B., Heald, C. L., & Hein, G. E. (1991). Increased exhibit accessibility through multisensory interaction. *Curator: The Museum Journal*, *34*(4), 273–290. https://doi.org/10.1111/j.2151-6952.1991.tb01473.x

Learning in Museums | Harvard Graduate School of Education. (2005, September 4). https://www.gse.harvard.edu/ideas/usable-knowledge/05/09/learning-museums

McNeil, T. J. (2023). *The exhibition and experience design handbook*. American Alliance of Museums.

Ramadan, M. E. (2024). Beyond Western perspectives: Inclusive insights to decolonise and transform science communication. *Journal of Science Communication*. SISSA Medialab srl. https://jcom.sissa.it/article/pubid/JCOM_2301_2024_R04/

REFERENCES

Allen, S. (2004). Designs for learning: Studying science museum exhibits that do more than entertain. *Science Education*, *88*(S1), S17–S33. https://doi.org/10.1002/sce.20016

Allen, S., & Gutwill, J. (2004). *Designing science museum exhibits with multiple interactive features: Five common pitfalls* (pp. 2–19). California Academy of Sciences.

Chien, H., Gutwill, J., Nee, J., & Shepard, J. (2025). Cultivating confidence: The potential of science museum visits in boosting science self-efficacy and expanding perceptions of science in emerging adult learners. *Science Education*, sce.21958. https://doi.org/10.1002/sce.21958

Evans, G. W. (1995). Learning and the physical environment. *Public institutions for personal learning: Establishing a research agenda*, 119–126.

Exhiblets – Free Science Workshop. (n.d.). Retrieved October 30, 2025, from https://freescienceworkshop.org/exhiblets/

Exploratorium. (2024). Exhiblets: A Guide for creating museum style exhibits in community spaces. https://www.exploratorium.edu/diy-exhibits

Hedge, A. (1995). Human-factor considerations in the design of museums to optimize their impact on learning. *Public institutions for personal learning: Establishing a research agenda*, 105–117.

McLean, K. (1993). *Planning for people in museum exhibitions*. Association of Science-Technology Centers.

Nee, J. (2024). *What do visitors learn from interacting with Exhiblets?* Exploratorium.

Rasekoala, E., & Orthia, L. (2020). Anti-racist science communication starts with recognising its globally diverse historical footprint. *Impact of Social Sciences Blog*. https://eprints.lse.ac.uk/105994/

Seeing | Exploratorium. (n.d.). Retrieved July 25, 2025, from https://www.exploratorium.edu/exhibit-collection/seeing

Serrell, B. (2015). *Exhibit labels: An interpretive approach*. Rowman & Littlefield Publishers, Incorporated. https://ebookcentral.proquest.com/lib/stanford-ebooks/detail.action?docID=2055684

Whale on Wheels | Mission Science. (n.d.). Retrieved October 30, 2025, from https://www.missionscienceworkshop.org/whale-on-wheels

CHAPTER 6

Scientific Opinion Writing

WHAT YOU'LL LEARN IN THIS CHAPTER

- An overview of the op-ed genre as it appears in popular media, including legacy newspapers, digital publications, scientific magazines, as well as the public-facing opinion sections of academic publications.
- An exploration of the two journalistic modes in which scientists choose to write in opinion genres: weighing in on *exigent* public conversations from the perspective of scientific expertise, and raising awareness about research and problems discovered by scientists in *evergreen* opinion articles.
- A taxonomy of several common situations in which scientists might choose or be asked to write an opinion piece: to *sound the alarm*, to *deliberate policy*, to offer *skepticism*, and to provide a *solution*.
- A roadmap for how op-eds establish credibility with diverse audiences and convey reliable, well-sourced scientific information, even when operating in spaces of polarization or uncertainty.
- Case studies that provide an account of the risks and benefits of taking a stand as a scientist.

DOI: 10.1201/9781003520887-8

- A process for developing an op-ed, from establishing your opinion-writing voice to strategizing how to use op-eds as a proving ground for emergent concepts, methods, and terminology.

INTRODUCTION

Giving your opinion in public, even as an expert, is always a risk. That's because changing one's opinion in public isn't always easy, even when it's based on the same proven principles and practices. For scientists and science communicators in the digital age, where an opinion could go viral or be cited detached from its original context, the risk may extend not only to one's personal reputation, but to the reputation of science at large. Yet risk is also an inherent part of being a scientist or science communicator. It's worth harnessing: insofar as research goals and agendas—and, for that matter, bias—always shape scientific discourse, the relationships between *research* and *opinion*, and between *research* and *activism*, are inherently fluid. Opinion writing, as a scientist, can take the form of research's cutting edge, where the inherently speculative quality of the opinion space can become a forum for trying out potential hypotheses and applications of research.

When a scientist chooses to give their opinion by summarizing the research in support of a course of action, there may be occasions where existing research directly explains an event or probable outcome; for instance, an epidemiologist sounding the alarm about an outbreak based on the known behavior of a disease. But there may be other situations where a scientist chooses to, or may be asked to, make an informed "guess." One might, for example, imagine the same epidemiologist trying to extrapolate lessons from previous epidemics to advise public health authorities on how to control a new, emergent disease. Scientific opinion writing can also vary in its urgency, from an *exigent* opinion in the context of a crisis, such as predicting epidemic activity, to what journalists call *evergreen* content, like speculation over whether humans should travel to Mars. Yet all of these scenarios contain an element of risk, because a scientific opinion is, by definition, a perspective, prediction, or judgment—and *not* a fact.

The publications in which you might publish an op-ed or other opinion genres also vary considerably in terms of goals and audience. A scientific op-ed in a legacy newspaper, such as *The New York Times*, is inherently different from a Viewpoint published in the *Journal of the American Medical Association*. That's because content is chosen and curated by editorial teams with different priorities, not to mention different readerships.

A newspaper or magazine will likely be trying to reach the broadest audience possible within a given geographical, political, or socioeconomic domain—and, in a digital age of audience editors prioritizing algorithmic pickup the most clicks. A *JAMA* Viewpoint, by contrast, speaks to, and in some cases for, readers of the journal, in this case primarily other researchers and physicians. However, because these pieces are more accessibly written, with less specialized jargon, they have the potential to reach larger audiences through citation.

Yet all of these rhetorical situations, despite their differences in timing and audience, share an inherent *speculative* quality that attends opinion genres, which makes them susceptible to attention, impact, and controversy that differ fundamentally from research and informational genres. Scientific opinion writing is riskier than scholarship and, for that matter, other public-facing genres, but it is also more widely distributed and read—which is exactly why you might consider writing an op-ed as part of your science communication portfolio. Opinion content is also especially promoted by publications, authors, and readers in social media spaces, where op-eds and other opinion forms may take on the appearance of reporting or even research, and where they might flounder or go viral. The speculative quality of op-ed writing, combined with the relative brevity of opinion content compared to most research genres, makes the op-ed especially available for popular consumption. It also makes opinion writing an especially compelling genre to study from the perspective of authorship and the writing process.

This chapter explores the opinion genres scientists write in, with a focus on the "op-ed," as well as the ethics and politics of when scientists should (and maybe shouldn't) weigh in on the things we care about. We'll look at both *exigent* and *evergreen* content, as well as several specific occasions in which opinion writing frequently transpires: to *deliberate policy*, to offer *skepticism*, and to *sound the alarm*. We will propose strategies for how to draft an op-ed, as well as what a writer can expect during the editorial process. We'll also explore potential pitfalls and reflect on strategies that might help mitigate the risk of misunderstanding and misrepresentation in digital spaces.

OPINION WRITING GENRES

Op-eds are usually relatively short. They generally range from 800 to 1,200 words. In digital formats, op-eds are also frequently written in easily digestible, 2–3-sentence paragraphs. Blocks of text in a digital op-ed are brief by design so they can be read within a single frame on a smartphone screen, and shared easily as social media posts. Another common feature of op-eds, especially those published in newspapers, is that authors generally do not choose their own headlines. In digital publishing, titles are usually written and sometimes updated by audience editors to optimize engagement through search engines, social media, and other content-distribution algorithms. In many cases, the lede (or intro) paragraphs of an op-ed might also be written by a newspaper's editorial staff to amplify an opinion's exigence, although authors should retain final sign-off before signing a contract.

Several distinct genres fall under the umbrella of scientific opinion writing. The most common opinion genre a scientist might find themselves writing in is an "op-ed," a newsroom and later popular shorthand that emerged in the first decades of the 20th century in the United States to describe the relatively short articles that appeared on the page "opposite" the "editorial" page of a print newspaper (*OED*). Whereas newspaper or magazine editorial pages, whether in their traditional print or 21st-century digital forms, refer to opinions given by the editorial staff of a publication itself, op-eds are usually submitted by or solicited from members of the public, although some publications, including major newspapers, employ a staff of regular opinion columnists with varied forms of expertise. Both editorial and op-ed content still exist under these labels in today's digital-first media landscape, although the placement of such content on digital news websites ranges widely depending upon publication and often changes depending upon the editorial priorities of a given publication. Most legacy newspaper websites still retain a dedicated opinion section to differentiate it from news reporting. But the lines can be blurrier on other news platforms: on the news website of a cable television channel, such as *CNN* or *Fox News*, content labeled as "opinion" or "op-ed" is far less organized, often blurring with reporting, which can be a deliberate strategy. Other kinds of publications, especially specialized science magazines, might have their own branded names for opinion content. The digital magazine *STAT*, which reports on medicine, biotechnology, and public health, refers to its opinion section as "First Opinion," a clever play on the idea of getting a "second opinion" from a doctor. In *The New Yorker* magazine, opinion content is labeled "Commentary." Some academic journals also have dedicated space for op-eds. The *Journal of the American Medical Association (JAMA)*, for instance, publishes "Viewpoints" from the medical community, which are similar in length and reader accessibility to newspaper op-eds, but which are published alongside research rather than news reporting.

Although the primary focus of this chapter is the op-ed genre itself, there are several related genres that are also relevant to the discussion. As a scientist or science communicator, you might have occasion to write a "letter to the editor," which also expresses an opinion, and which, depending on the publication, might be published alongside op-eds and editorial content.

Letters to the editor usually offer a response to published content and might constitute a critique, rebuttal, correction, or clarification, and many of the strategies relevant to op-ed writing are relevant here as well. A scientist might also be asked to give their opinion in an audio or video format, whether that takes place on a podcast, radio show, television interview, or other web-video format. Live-recorded audio and video formats are inherently more ephemeral than written contexts, but considerations regarding when and how scientists should weigh in with their opinions also apply. In the context of social or political controversies, a scientist might also coauthor or sign a petition, letter, or declaration, such as the Great Barrington Declaration or John Snow Memorandum, documents that were penned and circulated by scientists to influence government public health policy during the coronavirus pandemic. As noted above, we will also consider scientific opinion as a subcategory of scientific discourse—that is to say, how scientists communicate in all genres, including research—by thinking about the relationship between research, opinion, and activism.

THE OP-ED AND ITS MODES

Broadly speaking, there are two kinds of op-ed that you might consider writing as a science communicator: *exigent* and *evergreen*. The difference has to do with urgency, both in writing your op-ed and seeing it through to publication. Exigent content responds to a concrete current or future event, or another framing context that necessitates comment. If you feel like you want to give your opinion on a timeline that could run out or expire, then you are likely writing an exigent op-ed. An example of an exigent op-ed would be an epidemiologist weighing in on the probable activity of an emergent virus: that's an opinion other scientists and the public need as quickly as possible. Another example might be a climate scientist weighing in on an upcoming vote about a new coastal wind farm. Evergreen content, by contrast, is a term journalists use to refer to content not attended by the same urgency. A good way to tell if something is evergreen is to ask yourself whether an op-ed could reasonably be published at any time within the foreseeable future (with the caveat that the "foreseeable future" could always change). An evergreen op-ed might be, for instance, an evolutionary biologist arguing that the survival of the human species depends upon developing colonies in space, or a computer scientist arguing that investment in quantum computing is the only way for a national technology industry to retain a competitive edge.

Of course, what qualifies as "exigent" or "evergreen" is itself debatable, and, for that matter, determined by values, including values that may shift with changing political contexts, and so these two categories are perhaps best thought of as a spectrum. Yet an important distinction between these two kinds of content emerges in the publication process. An exigent op-ed might need to be written, edited, and published relatively quickly to remain relevant. A prewritten evergreen article or position, by contrast, might be written long in advance of publication. In some cases, an evergreen piece may be provisionally accepted by a news publication and then held unpublished, until a context emerges that would increase its audience and reach.

A final note: news sources inevitably tend to prioritize relevant, exigent op-eds on their opinion pages. This creates a further consideration for writers of op-eds, who, if they catch the ear of an editor, might feel pressured to find ways to make an evergreen piece fit into a current exigence.

STRATEGIES AND PITFALLS IN OP-ED WRITING: THREE CASE STUDIES

In the following section, we detail three example situations—*deliberating policy*, *offering skepticism*, and *sounding the alarm*—through case studies, exploring the specific rhetorical considerations of each scenario as well as potential pitfalls.

Deliberating Policy

One situation in which a scientist might want to give their opinion in public is when deliberating policy, whether that takes place at the global, national, state, or local level. Perhaps the greatest laboratory for this form of opinion writing in recent memory was the COVID-19 pandemic, when practices such as social distancing and masking became public mandates and subjects of intense debate and, sometimes, controversy. Take, for instance, the example of Dr. Monica Gandhi, who became an influential voice in US public health policy during the crisis. Gandhi is the director of the Gladstone Center for AIDS Research at the University of California,

San Francisco, and the medical director of the San Francisco General Hospital Ward 86 HIV Clinic. Her research focuses on women and HIV, and, in her public health work, she is a major advocate for harm-reduction and empowerment approaches to the treatment of diseases associated with social stigma. Deploying this perspective from her HIV work amid COVID, she was one of the earliest public health experts to call for universal masking (Gandhi & Havlir, 2020) and, sometimes in collaboration with aerosol scientist Linsey Marr at Virginia Tech, published opinion and research advocating for the use of better masks (Gandhi & Marr, 2020).

Later in the pandemic, on the basis of the same harm-reduction approaches to public health, once highly effective vaccines were developed, Gandhi became an advocate for off-ramps from interventions, specifically arguing that once people were vaccinated, they should be encouraged to return to normal life. In this context, she penned opinion pieces arguing specifically that schools should reopen once vaccines became available to avoid collateral harms to children (2021). The timing of Gandhi's shift in perspective complicated its reception, as initial perceptions that the vaccines would sterilize transmission in addition to preventing severe disease proved premature. Yet while she adamantly apologized for her mistake in timing, the public didn't forgive her easily. As science journalist Liz Highleyman points out, this created a whiplash effect in terms of her public reception: within the space of a few months, Gandhi, who had been derisively called "the mask queen" by critics of nonpharmaceutical interventions for COVID (mainly on the right), would later be named "the pandemic's wrongest woman" from the left, by MSNBC commentator Medhi Hassan (2023).

Gandhi's experience becomes a cautionary tale with several instructive lessons. It's significant, for one, how giving an opinion based on the same sound public health principles and considerable expertise could receive backlash from such radically different political camps, and for that matter, become a betrayal to some audiences—those who had relied on Gandhi as a voice of caution, in particular—when her opinion changed. The problem is that revising one's opinion is difficult in public-facing contexts, even when backed up by reliable research, evidence, and sound techniques. The timing of one's opinion, too, matters greatly, and amid moments of confusion and fear, there is a special risk of speaking too soon, before all the facts are known. Gandhi published op-eds and conducted research backing up her perspective, but she also offered some of her most candid commentary in interviews. Opinion writing is often meant to be provocative, and when rendered or repeated in viral videos and soundbites—taken out of context—it runs a special risk of amplification. In Gandhi's case, much of the backlash—from the left as well as the right—also sometimes took on a disturbing, misogynistic tone, underscoring the risk of exposure to attacks on personal identity in response to giving one's opinion.

A core lesson from Gandhi's experience is that deliberating policy as it depends upon or applies to science is inherently political. Op-eds that deliberate policy call attention to how science is political and are among the most explicitly political moves a scientist can make. Gandhi—who during the pandemic was active and widely followed on the social media platform then known as Twitter—was clearly aware of the inherent risk. As a tenured professor, her job was never in peril. Yet the experience was nothing if not fraught, and she recounts and explains this period in a subsequent book, *Endemic: A Post-Pandemic Playbook* (2023). That book—part memoir, part research paper, part detailed policy plan for future pandemics—details her rationale for taking the stands she did and when, including on topics such as masking and school closures. Citationally, this work is extremely dense: as dense as any formal scientific research paper. It also looks at the big picture, for instance in this moment where Gandhi reflects that her retrospective position is now one of trying to recover lessons for future policy deliberations: "We can learn, from HIV and now COVID-19, about compassion, reason, biomedical advances, transparency, equitable access, global development, and the importance of acknowledging uncertainty" (2023, p. 11). These are admirable, indeed evergreen, goals; however, like corrections to the scientific record as they get reported in the media, it bears noting that Gandhi's subsequent reflections on the story of her changing scientific opinion have not received nearly as much attention as her exigent claims did amid the COVID-19 pandemic. Scientists must always weigh what might be gained from taking a stand against the risks to themselves and to science that sensational attention to a claim might precipitate.

Offering Skepticism

Another powerful exigence to which a scientist might respond is the arrival of a new application of science—for instance, a new technique facilitated by the arrival of a new technology—which one would like to question, or with which one might even quarrel. A scientist might choose to write an op-ed offering skepticism because they feel themselves pulled into a conversation about which they have direct or proximal expertise, or where they feel motivated to say, *not so fast.* While skepticism might sometimes appear to some like the "brakes" on innovation, it's an important part of the public deliberation of policy and progress as they relate to science.

When offering skepticism, one might be tempted to take on a direct, contrarian stance. But a more rhetorically effective way of offering skepticism—even very harsh skepticism—can be achieved under the rubric of a question. Take the example of the opinion writing of John P. A. Ioannidis, for instance, a physician, meta-scientist, and statistician at Stanford University, who was the first person to publicly question the now-defunct biotechnology startup Theranos in a *JAMA* Viewpoint (2015). Although John Carreyrou would become famous in October 2015 for his investigative report in *The Wall Street Journal* debunking Theranos's claims that it could reinvent blood testing, Ioannidis' February 2015 *Viewpoint* opened the conversation with a simple question, contained in the title of the piece: "Stealth Research: Is Biomedical Innovation Happening Outside the Peer Reviewed Literature?"

Now, Ioannidis would be the first to admit that peer-review is an imperfect process, subject to bias and error. After all, he became famous and built his reputation on an article that is widely considered to be one of the most influential and widely cited of all time: "Why Most Published Research Findings Are False" (2005). But he had a hunch that Theranos' lack of transparency when it came to its "black-box" blood-testing technology signaled larger problems within the company, a hunch that turned out to be correct.

In the op-ed, Ioannidis is rigorous in his questioning, but careful not to prematurely accuse Theranos of deception. To walk through a few of his key moves, he first works to map out the broader scientific publishing and innovation landscape in which the company's claims emerged. In the Viewpoint, Theranos is "just one example among many" biomedical firms offering findings with a scientific veneer—or a "biotechnology flavor," as Ioannidis puts it—but which ultimately intend to offer "products and services" rather than advances in scholarship (2015, p. 661). The "spirit of private ultracompetitive entrepreneurship" pervades here, rather than the accountability and transparency inherent in academic research (p. 661). Although Ioannidis presents this as a problem, what's interesting is that he generalizes his claims, leveling them not at Theranos directly, but at a broader trend of "stealth research" that arrives amid sensationalist advertising, press releases, and media claims. The key question, for Ioannidis, is this: "How can the validity of the claims made be assessed, if the evidence is not within reach of other scientists to evaluate and scrutinize?" (p. 661).

Bear in mind that Ioannidis ultimately *does* want to single out Theranos here: he has (at least) a hunch that something is amiss within the company. Yet he is careful to do so in a balanced way, citing it as but one notable expression of a larger, potentially problematic trend in innovation. By giving the company the benefit of the doubt, even amid his intense skepticism, he builds ethos with the reader as a fair evaluator of the situation. Notice his care in choosing language to describe the situation as potentially problematic, without blatantly indicting the company. Their secrecy is not necessarily the mark of corruption (although it *would* turn out to be this), but merely "paradoxical," given other trends at the time in laboratory and clinical research toward increasing transparency. His recognition, too, that peer reviewers can be biased against those claiming to offer an innovation also cautiously balances his skepticism. Ioannidis also introduces a portable, concrete term for the phenomenon he attempts to call out here, naming what Theranos and other biotechnology companies might be up to as "stealth research." This would be a term Ioannidis would return to in subsequent research, including in the 2019 peer-reviewed study "Stealth Research: Lack of Peer-Reviewed Evidence From Healthcare Unicorns," which he co-wrote after Theranos' implosion (Cristea et al., 2019), demonstrating how ideas and terminology prototyped in the context of an opinion article can—once vindicated—be redeployed and elaborated in the context of subsequent peer-reviewed research.

Calling for better and more transparent evidence is the modus operandi of Ioannidis and his Stanford team. Yet despite being correct about Theranos, Ioannidis is not infallible as a cautiously skeptical voice in science. When Ioannidis sought to question public health interventions during the extreme fear and polarization of the early days of the COVID-19 pandemic in the same mode—calling for better evidence—the gesture took on a very different valence. In March 2020, writing a "First Opinion" for the health and medicine news website *STAT*, he made the claim that the data underwriting large-scale public health interventions in COVID policy, most notably large-scale lockdowns and stay-at-home orders, were insufficient. Although the claim at the basis of his argument was sound—we did *not* have reliable data during the early days of the pandemic, because of insufficient testing and sample size—the response to his op-ed went haywire for several instructive reasons. The article's title—which echoes the questioning rhetorical stance of the *JAMA* Theranos op-ed—might seem barely controversial today, except insofar as it deviated from support for the many interventions of the early days of COVID: "A fiasco in the making? As the coronavirus pandemic takes hold, we are making decisions without reliable data." In the article, he explores a range of hypotheticals, arguing that the infection fatality rate from COVID—in March 2020—was not yet known, and that the disease may have been either far more dangerous or far less dangerous than could be supposed at the time. The problem, as it was in the case of Theranos, was a lack of reliable evidence, and the scientist was mainly advocating for accelerated and expanded testing, warning that long-term, disruptive interventions, however well intentioned, could have unintended negative consequences. This time, the supercharged climate of fear and political polarization of the early days of the pandemic, as well as a widely cited metaphor Ioannidis used to describe the lower-fatality-risk possibility, caused the article to go viral on what was then still Twitter. Ioannidis wrote, of potential overreaction to a possible best-case scenario of the virus' case-fatality rate, "It's like an elephant being attacked by a house cat. Frustrated and trying to avoid the cat, the elephant accidentally jumps off a cliff and dies." The *STAT* piece was rapidly written up on conservative firebrand Ben Shapiro's website, *The Daily Wire* (Bickley 2020), which was then promoted by Elon Musk (2020) on social media amid his highly vocal calls for the immediate reopening of Tesla factories.

No one was more frustrated by this misrepresentation than Ioannidis himself, who expressed in many public contexts his dismay over statements citing lines from his op-ed out of context and has broadly reiterated his support for caution during the early days of the pandemic (Prasad, 2022). But in the polarized climate in which he was writing, the metaphor he deployed for the possibility that the virus might be less deadly than supposed, which was largely the vehicle of his opinion's virality, took on a life of its own.

Now, let's just observe that this is a somewhat unstable metaphor. Why are the cat and the elephant hanging out by a cliff? Do we have any idea? Clearly this imagined scene is meant to capture our attention and be memorable, which in a different context might be an admirable quality in a metaphor deployed in an op-ed (or in another genre where metaphors are useful in science communication, such as podcasts, museum exhibits, and explainer videos). Metaphors—even slightly bonkers metaphors—can be useful tools in conveying science (see Chapter 2, *Rhetorical Tools*, for an in-depth account of metaphors and their uses, as well as specific risks). But herein lies the danger, too, and the core lesson for us as science communicators: our opinions, and just as importantly the language with which we choose to present them—including our metaphors—are subject to wild, unruly proliferation outside the opinion genres where they might seem safely speculative *because* of the special contiguity between public-facing op-ed genres and social media, as well as traditional mass media.

Risk indeed: the proliferating entanglements here are stunning and should give all scientists pause before wading into the waters of scientific opinion. An important takeaway in this context is how—not unlike Monica Gandhi—Ioannidis took on significant professional and personal risk by offering his reasonable skepticism in a politically supercharged environment. Yet the most instructive lesson of Ioannidis' experience is about language, and how soundbites from one's argument—in this case, a vivid but unfortunately imprecise metaphor—can circulate detached from context regardless of the soundness of one's argument or intentions.

Sounding the Alarm

Sounding the alarm poses a special challenge as an exigent opinion mode, given that the idea of providing advance warning necessarily implies that many people—or even most people—may not be aware of the dangers you seek to warn them about. The risks to professional and scientific credibility are high, insofar as a scientist warning of a new or unwelcome danger could be accused of "crying wolf." Yet scientists, largely because of their specialized knowledge and position on the cutting edge of research, are often the ones who find themselves in the position of having to sound the alarm. When they choose to accept that responsibility, it serves an important societal role, because it's important for warnings to be backed up by expertise.

Sounding the alarm differs from deliberating policy because its purpose is to *start a broader conversation*, rather than debate nuances of policy. While sounding the alarm may challenge or undermine a dominant conversation or trend, it also differs from what we define in this chapter as skepticism insofar as it is less granular in its critique. When you sound the alarm, you offer a significant reevaluation of a situation based on a new or birds-eye-view, for instance, with systems knowledge. Such calls, when made effectively, are usually attended by a specific agenda for further study, discussion, or future action.

The trick to effectively sounding the alarm is balancing the urgency of one's tone with the practical implications of one's message. Writers in this submode might find it productive to look for opportunities to build upon examples of topics about which there is already public concern. Take, for instance, Linsey Marr's June 2023 op-ed for *The New York Times* raising the alarm about indoor air quality, "What Wildfire Smoke, Gas Stoves and Covid Tell Us About Our Air." Marr—a sometime collaborator with Monica Gandhi cited above–is an engineer who specializes in aerosol science at Virginia Tech and, like some of the other authors we've discussed in this chapter, became a significant voice during the COVID-19 pandemic, in her case once it became clear that the virus was airborne and her specialized expertise in aerosols would be useful. While her title in this instance would have been developed in collaboration with audience editors at the *Times*, it captures Marr's basic move in the article: to combine three separate concerns regarding air quality into a broader call for increased attention to air quality standards in *all* spaces. While governments routinely report on outdoor air quality, especially during intense pollution and smoke events, indoor standards are barely regulated. Marr's opening gambit starts by citing the immediate exigence of the wildfires that were blanketing much of North America in the summer of 2023. But right away, she uses this exigence to pivot to a reason why we should be rethinking indoor air quality now:

> If the pandemic was whispering to us about air quality, the wildfires are screaming to us about it. Add to that concerns about gas stoves and longer allergy seasons, and it's clear we should be on the precipice of a new public health movement to improve the air we breathe.

What's powerful about this approach is how it layers several existing reasons for caring about air quality into a new, refocused, generalized call. This allows the author to appeal to a potentially broader audience than she would have had she focused more narrowly on a single factor. In the summer of 2023, wildfires in Canada spread an alarming layer of smoke across much of North America. Around this same time, gas stoves were in the news, with multiple jurisdictions in the United States considering banning indoor gas stoves and furnaces because of their impact on air quality. While both of these topics (not to mention COVID-19) were politicized in some spaces, citing them here as an accumulation of related potential reasons to rethink the importance of air quality both stays out of the "weeds" of any of these specific questions and appeals to parties who might have a special interest in one of them. Like Ioannidis in his Theranos essay, this op-ed offers its primary suggestion as a question rather than a directive, asking why governments are not doing more to regulate indoor as well as outdoor air quality indoors. Although Marr indeed calls for significant policy change, by framing her suggestion as a question, she invites the audience to join her in a collaborative process of rethinking the importance of indoor air quality.

GETTING YOUR OP-ED OFF THE GROUND

Once you are ready to break ground on the op-ed itself, there are a few conventions of the genre that warrant special consideration. Foremost among these is the relative brevity of op-eds, and the expectation that most will be no more than 800–1,200 words in length. How do you keep it short, especially if you are used to the relative freedom of longer forms, such as an IMRD (Introduction-Method-Results-Discussion) research paper?

What's important to remember is that even though a published op-ed will be relatively short, op-eds are often the product of many revisions. If your initial draft is long, there's nothing to worry about, but it might provide an occasion to rethink the scope of your argument and topic. It could be helpful to ask yourself: *Am I actually writing about multiple topics that need to be separated, or do I need to distill my ideas?* One could hypothetically imagine, for instance, that Linsey Marr (cited above) might have written three separate op-eds on different aspects of air quality, and as an aerosol scientist she very likely has more to say individually about the problems of wildfire smoke, viruses in the air, and appliance exhaust. But the point of her article is that these three issues come together to create a distilled argument for rethinking air quality in general as a public health problem.

Because of the relative brevity of op-eds, as well as the likelihood that they may be shaped by editors looking to capture the exigence of a moment, the ideation and drafting process for op-eds can look somewhat different from that of other science writing genres. It's important to remember that the short length of op-eds does not necessarily mean that they take less time to write: effective op-eds are not only concise, they are also dense and nimble, moving rapidly between claims, provocations, evidence, and examples. These features of op-eds mean they can actually take a lot of time to craft. Effective op-eds often display significant citational density: although you will probably not be required to generate a bibliography or works-cited list accompanying your op-ed, depending upon the publication, digital content will usually cite its sources and reference points via hyperlink.

Perhaps more than other genres in this book, op-eds are usually occasional pieces; that is to say, they are not usually part of a series, and the occasion for writing them usually emerges because the author has either been asked to write one or has found an urgent exigence to which they'd like to respond. A scientist might write a dozen op-eds during a time of crisis, like the COVID-19 pandemic, and then return to research and other genres.

If you find yourself with an occasion to write an op-ed, however, and think you have an idea you want to weigh in on and are ready to start writing, here are some generative questions to help you shape your draft:

- What gives you the credibility to speak on this topic? Do you have special expertise with on the topic? Personal experience?
- What establishes the exigence for your op-ed? What current events or situations warrant your proposed commentary, solution, or intervention?
- What audiences do you want to reach with your op-ed?
- What examples, representative cases, and metaphors might help your audience understand and care about your position?
- How does your op-ed relate to and mutually benefit from or reinforce your work in other genres? What ideas might you want to "try out" publicly in the context of an op-ed that might become a component of subsequent research?

And a few cautionary questions, to keep things on the rails:

- Do any of the examples, cases, and/or metaphors you deploy run the risk of being circulated out of context?
- What personal and professional risks do you take by taking a stand in public in the context of an op-ed? Are you comfortable taking these risks?

PITCHING AND THE EDITORIAL PROCESS

Although scientists who have specialized expertise or established platforms may be approached by publications to weigh in on a topic of public concern, for most science communicators, if you have an idea for an op-ed, you'll need to submit it to a publication. This will probably involve writing a pitch. In some cases, a publication may ask for a draft of the complete op-ed as well, but you'll still need to write a pitch framing your argument, most likely in an email. Because publications cater to specific audiences and may have specific length requirements and editorial procedures, it can often be helpful to identify a possible publication venue and send out your pitch *before* drafting your op-ed. In fact, writing your pitch as a kind of "abstract" for your op-ed can be a way to get the project off the ground. Aspects of your pitch might be imported directly into the draft of the op-ed itself.

So, what makes an effective pitch? Pitches are usually sent via email to a publication's opinion editor and foreground, in a few sentences, the following:

1. Your central claim, argument, or intervention.
2. The exigence and context for why your op-ed matters.
3. Why the audience of a given publication would be a good fit for your op-ed.

In your pitch, you should also provide a little relevant information about yourself, including your experience and credentials, as well as any prior publications. If you have a personal or professional website, or a social media presence relevant to your op-ed, you should certainly include links to those items as well.

You should address your pitch as a formal letter, but don't be afraid to make it interesting or even provocative. If there is a proximate current event that makes the topic exigent, for instance, make that clear in your pitch. You want an editor to be as excited as you are about the possibility of publishing your op-ed, and to understand how it will speak to their audience. If you don't hear back within a week, it's usually acceptable to send a follow-up email to ask whether the editorial team has considered your pitch, although you should always make sure to double-check regarding response timeline specifications on the publication's submissions page. If a publication explicitly states that it can take up to four weeks for a response, for instance, it's best not to risk pestering a busy editor prematurely.

Usually, when a pitch or draft op-ed is accepted, it will undergo an editorial process, which can take time. If you're working with a newspaper, for example, a paper's opinion editor might express interest in your idea but will still be thinking about a broader schedule for the paper's content, given that only so many op-eds can be published in a given issue. Urgent topics might get triaged to the front of the line by a newspaper, but, following initial acceptance, more evergreen op-eds might wait in a queue for weeks or even months until the right moment presents itself. It's also typical for newspapers to suggest titles to promote engagement and even to rewrite the entire lede of a piece in order to shape its relevance. As an author, remember that you are ultimately on the line and in control of these decisions: if a publication offers a dismissively long timeline to publish an op-ed that you think is more urgent, or makes edits that you don't agree with, you can always take your content elsewhere until you sign an author contract.

Pitching Exercise: Practicing an Exigent Pitch

Open up the homepage of a major scientific journal—for instance, *Nature* or *Science*—and take a look at its recent news briefs, where journals often provide public-facing overviews of emerging research. Without necessarily knowing your final position, imagine a pitch that you could send to the editor of one of the publications on the list below for a possible op-ed. In 150–200 words maximum, draft a concise, focused, and interesting pitch email. How might you get the editor's attention? What specific items, questions, and/or buzzwords might the publication in question be looking for? Who is the publication's audience, and what aspects of the research might be most interesting to them? What comparable studies, methods, or situations might you use by analogy to get your idea off the ground with the editor?

Possible op-ed publication venues: *The New York Times, The Atlantic, Undark, MIT Technology Review, Wired, Vice, Scientific American,* a local newspaper in your area, a college newspaper, or another publication of your choice.

REFERENCES

Bhattacharya, J., Gupta, S., & Kulldorff, M. (2020, October). *Great Barrington Declaration and Petition*. gbdeclaration.org/

Bickley, J. (2020, March). Stanford professor: Data indicates we're severely overreacting to coronavirus. *The Daily Wire*. dailywire.com/news/stanford-professor-data-indicates-were-overreacting-to-coronavirus.

Carreyrou, J. (2015, October). Hot startup Theranos has struggled with its blood-test technology. *The Wall Street Journal*. wsj.com/articles/theranos-has-struggled-with-blood-tests-1444881901?reflink=desktopwebshare_permalink

Cristea, I., Cahan, E., & Ioannidis, J. P. A. (2019). Stealth research: lack of peer-reviewed evidence from healthcare unicorns. *European Journal of Clinical Investigation.* doi.org/10.1111/eci.13072

Gandhi, M. (2021, February). Overcaution carries its own danger to children. *The Atlantic.* theatlantic.com/ideas/archive/2021/02/vaccines-are-banishing-any-debate-about-reopening-schools/618155/

Gandhi, M. (2023). *Endemic: A post-pandemic playbook.* Mayo Clinic Press.

Gandhi, M., & Havlir, D. (2020, April). The time for universal masking of the public for coronavirus disease 2019 is now. *Open Forum Infectious Diseases.* doi.org/10.1093/ofid/ofaa131

Gandhi, M, & Marr, L. (2020). Uniting infectious disease and physical science principles on the importance of face masks for covid-19. *Med.* doi.org/10.1016/j.medj.2020.12.008

Highleyman, L. (2023, September). What the pandemic's 'wrongest woman' got right. *Slate.* slate.com/technology/2023/09/endemic-monica-gandhi-covid-masks-lockdowns-school-closures.html

Ioannidis, J. P. A. (2020, March). A fiasco in the making? As the coronavirus pandemic takes hold, we are making decisions without reliable data. *STAT.* 17 Mar. 2020. statnews.com/2020/03/17/a-fiasco-in-the-making-as-the-coronavirus-pandemic-takes-hold-we-are-making-decisions-without-reliable-data/. Accessed 16 Aug. 2025.

Ioannidis, J. P. A. (2021, December). Where are we now with covid-19? (interview with Vinay Prasad). *YouTube.* https://www.youtube.com/watch?v=YvEOwKUyur0

Ioannidis, J. P. A. (2005). Why most published research findings are false. *PLOS Medicine.* doi.org/10.1371/journal.pmed.0020124

Ioannidis, J. P. A. (2015). Stealth research: Is biomedical innovation happening outside the peer-reviewed literature? *JAMA.* doi.org/10.1001/jama.2014.17662.

Marr, L. (2023). What wildfire smoke, gas stoves, and covid tell us about our air. *The New York Times.* https://www.nytimes.com/2023/06/07/opinion/wildfire-smoke-air-quality-viruses.html.

Musk, E [@elonmusk]. (2020, March). Imo this professor is correct. *Twitter.* https://x.com/elonmusk/status/1240707187229380608

Prasad, V. (2022, January 4). Where are we now with Covid-19? With John Ioannidis (No. 4.36) [Audio podcast episode.] In *Plenary session.* https://www.plenarysessionpodcast.com/episodes/jw9w58yrajfyr6r-7b8sr-6xwsz-6m9zm-y9bgz-wkczr-rkbad-5knrb-cr2ah

N. Alwan, R. Burgess, S. Ashworth, R. Beale, N. Bhadelia, D. Bogaert, J. Dowd, I. Eckerle, L. Goldman, T. Greenhalgh, D. Gurdasani, A. Hamdy, W. Hanage, E. Hodcroft, Z. Hyde, P. Kellam, M. Kelly-Irving, F. Krammer, M.Lipsitsch,...H. Ziauddeen. (2020, October). Scientific consensus on the COVID-19 pandemic: we need to act now [The John Snow memorandum]. *The Lancet.* doi.org/10.1016/S0140-6736(20)32153-X

CHAPTER 7

Profiles

Makers of Science

WHAT YOU'LL LEARN IN THIS CHAPTER

- A definition of the profile genre
- How profiles of individuals can make scientific concepts and worlds accessible and compelling
- A process for writing your own profile, from subject selection through interviewing to crafting character-driven narratives

INTRODUCTION

Whereas research papers present findings through data and methodology, profiles invite us into the lived experiences of the people who make or are transformed by scientific knowledge. Profiles in the modern sense, as a genre of short- or long-form literary journalism, have appeared in the pages of *The New Yorker* since their late-1920s debut in that venerable American publication under the editorship of Harold Ross (Remnick, 2000). Influenced by a long tradition of biography—from writers as varied as the ancient philosopher Plutarch to the Enlightenment novelist Daniel Defoe—the profile feature in *The New Yorker* innovated "a concise rendering of a life through anecdote, incident, interview, and description" that continues to influence the form (Remnick, 2000). Still today, "intimacy and wit" are emphasized "over biographical completeness," in the

DOI: 10.1201/9781003520887-9

words of David Remnick, who edited an anthology of *New Yorker* profiles (Remnick, 2000).

Consider D.T. Max's 2017 *New Yorker* profile of Jim Simons, a mathematician who once led the math department at the State University of New York at Stony Brook and later became the billionaire founder of the hedge fund Renaissance Technologies. Rather than simply documenting Simons' career, Max shows us Simons as a character: affable, sometimes defensive, and at times so caught up in mathematical problems that he detaches from conversation. In other words, we encounter Simons on the page as fully human—flawed and controversial. Readers learn that Simons won the Veblen Prize, but that's not the point of the profile, which is interested instead in the bigger questions that his unique career raises at a particular moment in time, after he has found remarkable success in both science and business (Max, 2017). The piece begins in the offices of the Flatiron Institute, a scientific think tank fueling breakthroughs in basic science that Simons' fortune has funded. As it describes Simons at work leading this new institute, Max's profile asks whether the basic sciences can be reliably funded by private institutes.

This is the power of the profile in science communication. Profiles translate specialist debates and the complex social dynamics of science into grounded, lively human stories.

WHY SCIENCE COMMUNICATORS CHOOSE THE PROFILE

Profiles serve science communication in several crucial ways. Perhaps most important, they make the invisible visible. Much scientific work happens in laboratories, behind computer screens, or in remote field sites that most readers will never see. It can even happen at home and emerge from idiosyncratic interests. When *The Atlantic*'s Mark Bowden profiles computer science professor Larry Smarr in "The Measured Man," we are taken into Smarr's private home and daily routine. Bowden shows us the ways that Smarr meticulously documents his health, tracking hundreds of separate markers from his blood, skin, stool, and sleep. Smarr's work as a computer scientist at the University of California had taught him how to capture, parse, and store vast troves of data, an approach he felt could help individuals anticipate health problems before they become problems, while they're still "pre-clinical," something that, in his case, proves true. Through his extensive monitoring, Smarr discovers that he's developing Crohn's disease before his doctors do. The science becomes inseparable from the scientist.

Bowden relates Smarr's personal investigations to larger conversations in the health sciences about the value of big data collected from millions of individuals engaged in daily monitoring. While Smarr's example shows why the approach can be successful, it also reveals a downside: it can produce obsessions, which many medical doctors argue are counterproductive and undermine health, as Bowden is able to show through quotes from skeptical researchers (Bowden, 2012). In other words, profiles also humanize controversy, another reason science communicators may choose the profile genre. Scientific debates often appear as bloodless exchanges of data, but they're driven by real people with real stakes in the outcomes. Human context can make complex scientific disagreements more comprehensible and engaging for general audiences.

By showing scientists in their communities—their laboratories, their collaborations, their conflicts—profiles reveal that science isn't the work of isolated geniuses but of people embedded in institutions, cultures, and relationships. This perspective can help readers understand how scientific knowledge develops and changes over time: that it is always relational and negotiated within communities. Finally, science communicators might choose to write a profile to advocate for change by showing the human consequences of science. To take just one example, medical writers profile patients whose lives have been transformed by new treatments, helping readers understand what's at stake in medical research.

THE PROFILE GENRE: PROFILE VS. INTERVIEW VS. BIOGRAPHY

A profile is more than a character sketch or interview transcript. It's a carefully constructed narrative that uses one person's story to illuminate larger themes, issues, or worlds. The writer chooses which details to include, which moments to dramatize, and which aspects of a person's character to emphasize. It's partial, in other words, not a complete portrait, and thus, at its best, communicates a distinct perspective on the scientist. But this isn't a weakness; it's what allows profiles to be both informative and emotionally engaging.

Understanding the profile genre means distinguishing it from related forms: the interview and the biography (for more on being interviewed by the media as a scientist, see Chapter 9, "Presentations and Oral Communication"). An interview is typically structured as a question-and-answer exchange, with the interviewer's questions visible and the subject's responses given in their own words. It is almost always an edited and condensed version of a longer conversation between the science communicator and the scientist. Below, we recommend that you conduct at least one interview with your subject before you write up the profile as part of your research process, but here we are talking about the version of the genre that is polished and published. As an example, we point you to the archive of scientist interviews on the website of NASA, the National Aeronautics and Space Administration of the United States. In one instance, an interview with Caleb Scharf, an astrobiologist, meanders through his childhood, training, discoveries made through specific new telescopes, and advice for postdocs (Fitzpatrick, 2025). Thus, an interview may illuminate a scientist's work or a scientific controversy, but it won't be unified by a narrative.

By contrast—and at the other end of the spectrum when it comes to length—a biography is a comprehensive, book-length account of someone's entire life, usually chronological, which we might think of as an extended profile. Science communicator Seema Yasmin elected to write a biography of "the impatient" Dr. Joep Lange, the pioneering Dutch physician who fought to understand and treat HIV, the virus that causes acquired immuno deficiency syndrome (AIDS). The longer form allows Yasmin to situate Lange's work in the history of the global AIDS epidemic and to honor his advocacy for compassionate, prophylactic treatment; it also required a significant amount of research, as Yasmin weaves in content from medical journals, written testimony, and extensive interviews (Yasmin, 2018). Because it's also a lengthy print or audiobook, it is less accessible to the public than a compact profile.

A profile sits between the biography and the interview. Like an interview, a profile is generally based on direct conversation with the subject. But unlike an interview, the writer's questions usually disappear, replaced by narrative that weaves together the subject's words, the writer's observations, and background research. Like a biography, a profile tells a life story. But unlike a biography, it typically focuses on a specific period, challenge, or slice of that life. Consider the implication of the metaphor: in art or photography, a profile gives us just one view of the face—the view from the side. It's not a complete portrait, which is what biographies often strive to provide. The profile is thus typically much briefer than a biography, making it very versatile. You might hear a profile when a scientist is introduced before giving a talk. A university communications team might write profiles to help publicize the scientific work at their institution. And instructors often use the profile to make teaching about science engaging.

The profile may even be incorporated into another genre, as a subset of it—for example, into a podcast conversation with a scientist or into a long magazine story, as we see in Atul Gawande's 2007 profile of Dr. Felix Silverstone at the end of the 2007 *New Yorker* story "The Way We Age Now." Many profiles are just 1,000 words long, though profiles can also stretch into long-form essays—5,000 or 6,000 words—and include a great deal of scientific context as well as details about the subject's character and work.

Most importantly, profiles are built around scenes—moments when we see the subject doing something in a specific time and place. These scenes alternate with reflection, background information, or analysis. This structure allows profiles to function as both story and exposition, entertainment and education.

SETTING, CHARACTER, AND PLOT IN SCIENCE PROFILES

Even though a profile is nonfiction and based on fact, it deploys story elements that it shares with fiction. Every compelling profile contains three essential story elements—setting, character, and plot. In science communication, each element serves specific purposes. (For more on these elements of story, see Chapter 2 as well as Chapter 4 on personal narrative. Here, the elements are described in relation to the profile.)

Setting grounds the story in specific places and times, but it also reveals character and context. The cluttered laboratory, the sterile hospital room, the remote field station—these details aren't just background decoration but windows into the subject's world and priorities. In science profiles, the piece often begins with the setting because that's where the problem that opens the story first emerges. If you aren't sure where to start your piece, consider where your scientist works, whether at a computer monitor or on a ship in the open sea. Describe the setting as you describe the scientist's motive: the confusion, doubt, or question within the larger debate of the field that will pull your reader in and launch your story.

Character is more than physical description or biographical facts, though these elements are important. It's the subject's distinctive way of moving through the world. Pulitzer Prize–winning journalist Jacqui Banaszynski argues that profiling requires asking "the deep questions: What has defined this person? What is this person's motivation? Value system? Approach to life? Who *is* this person?" (Banaszynski, 2007). In science profiles, character emerges through the subject's relationship to their work—their curiosity, persistence, and particular way of seeing problems.

Plot is the sequence of events that reveals character and creates meaning —either the subject changes the world through their work, or the world changes them, or both. Science profiles often follow a researcher through a discovery process, a patient through a treatment journey, or an innovator through the development of a new technology. The key is finding moments of tension or transformation where something is at stake.

Consider how these elements work together in Benedict Carey's 2017 *New York Times* profile of neuroscientist Brenda Milner, "Eminent Brain Scientist Is 'Still Nosy' at 98." The profile opens in Milner's office; in fact, at the top of the written profile, we are presented with a photo of the scientist in her office, surrounded by towers of paper files representing decades of reading. This matters because, 3 days a week, she comes in to work in a dedicated space at McGill University, where she is a professor of psychology. Milner's character emerges through a number of modalities: her wry wit as she shares an anecdote, her careful attention to appearance, which Carey evokes through descriptions of her dress, and her longtime habit of meticulous observation in life as in science. The plot asks how she arrived at her new project, which Carey shows emerged from two important strands of psychology research, one of which she innovated. The first follows her work with patient H.M., as he is known in the research literature, whose brain surgery left him unable to form new memories, and what his case taught her about memory formation. The second addresses the field's attention to hemispheric specialization, "how the brain's two halves, the right and the left, divide up its mental labor," as Carey puts it (Carey, 2017). The plot of the profile takes us through the ways Milner found answers to open questions, often through new methods, leading to what we might think of as its climax. Though it's not dramatic in the way a climax might be in the genre of a thriller, the description of Milner's current project represents a logical culmination of her scientific work and curious personality. In this profile, character and action work together to communicate an important subfield of neuroscience.

SELECTING YOUR SUBJECT

The best profile subjects aren't necessarily the most famous scientists; rather, they're people whose stories illuminate something important about science or technology that your readers need to understand. In fact, part of your goal in composing a profile is to show why the subject is worthy of the news and your reader's attention. Even if their contribution to science is relatively small so far, their story can still provide insight into a field or the scientific process that you think deserves wider attention. So, your first step may be to reflect on which narrow area of science you want to communicate and why. Here are some approaches to selecting compelling subjects within that area:

- Whose work or experience represents a trend or issue? A young climate scientist observing sea-level rise might exemplify how early-career researchers are grappling with urgent environmental challenges, for example.
- Whose work is forging new methods or technologies? This might be a researcher developing new ways to measure whale calls or an engineer designing more sustainable batteries.
- Whose life has been shaped by scientific or medical developments? Patients with rare diseases, communities dealing with environmental changes, or astronauts navigating space all offer windows into how science and technology affect real lives.
- Who is at the center of scientific debates or ethical dilemmas? This approach requires especially careful handling but can illuminate important issues in policy that are strongly affected by science or research ethics.

These categories might overlap in the person you're interested in—if so, that just makes them more interesting! The key is eliciting the story that makes the scientist more interesting than even they may realize—not just general expertise, but a particular challenge they've faced, discovery they've made, or experience they've lived through that will resonate with your readers. Put more strongly, you don't need to hunt for someone with a story to tell; rather, *work to bring a story out of someone who interests both you and your audience.* Below, we suggest a research and writing process that includes an interview with your subject to help you do this.

TAKING CARE WITH YOUR CONTENT AND REPRESENTATION

Profiling raises important ethical questions, especially when dealing with scientists whose work affects public health or policy, or patients whose medical conditions make them vulnerable. The profile writer holds significant power in deciding how to represent their subject, and this power comes with responsibilities. Profiles aren't journalism in the strictest sense—they're often more impressionistic than investigative—but they still require factual accuracy about the science being discussed. If you're profiling a researcher, you need to understand their work well enough to represent it fairly, even when translating technical concepts for general audiences.

You should always obtain your subject's consent and respect their privacy. Unlike public figures, who expect media attention and may give frequent interviews, many scientists aren't accustomed to being profiled. Consent is even more important—and a prerequisite—when profiling patients and members of other vulnerable or potentially vulnerable groups. It's crucial to be transparent about your intentions, to respect the boundaries your subject sets, and to consider how your portrayal might affect their professional or personal life. In navigating the ethics of profiling vulnerable subjects in particular, you must consult the IRB (Institutional Review Board) at your university, if you are affiliated with one. You might also consult online statements such as the Principles of Healthcare Journalists from the Association of Healthcare Journalists that suggest important steps you can take to minimize harm while providing fair representations of the science.

Profiles can be powerful tools for advocacy, but they shouldn't manipulate readers' emotions or oversimplify complex issues. The goal is to illuminate, not to sensationalize.

CONCLUSION

The profile offers science communicators a powerful tool for making distant discoveries familiar. When profiles focus on those who make science, they invite us into the communities, institutions, and cultures that shape scientific knowledge. They show us that science is a human endeavor driven by curiosity and hope, ambition and fear. When profiles focus on those impacted by science, readers see how new knowledge solves problems and improves quality of life. Whether you're profiling a Nobel laureate or a first-year graduate student, a patient navigating experimental treatment or a community leader fighting environmental injustice, you show the public that we both make science and are changed by it.

A PROCESS FOR COMMUNICATING SCIENCE THROUGH PROFILES

Writing a strong profile requires careful preparation, skilled interviewing, and thoughtful narrative construction. The process typically unfolds in several stages.

Read Sample Interviews, Profiles, and Biographies Like a Writer

When we read "like a writer," we read with an eye and an ear for strategies that we might use in our own compositions. Ask yourself, what qualities of this text would I like to emulate (or avoid) in my own writing? Reading biographies, for example, can give you strategies for rendering character and situating a life in a larger context or controversy. Reading lots of profiles can then show you how to narrow your scope. *The New Yorker* profile archive mentioned at the opening of this chapter is a great place to start. And reading published interviews can help you brainstorm your own list of questions and understand what profiles provide readers that interviews alone cannot.

Research and Preparation

It's possible that you may not be able to speak with your subject. If so, you can still develop a profile based on what you're able to gather and infer from other sources, including their published work and what others have said about them. But a profile informed by an interview with the subject will have more depth and originality, so it is worth trying to make it happen.

Before contacting your subject, do your homework. If you're profiling a scientist, read their recent papers, become familiar with some of the foundational concepts in the field, and identify the specific aspects of their work you want to explore. If you're profiling a patient or community member, research the medical or environmental issues they're dealing with. This preparation serves two purposes: it helps you ask better questions, and it demonstrates respect for your subject's time and expertise.

Look for tension and turning points in your subject's story. Has there been a recent breakthrough or setback? A career change or personal challenge? A moment when the ground shifted? These inflection points often help narrow your story and, thus, the range of questions you will ask. Don't go into an interview with a busy cardiologist, for example, expecting to learn about the basics of catheterization. Instead, know which moment in the cardiologist's career most interests you and why.

When you email your subject to ask for an interview, describe your reasons for reaching out, which may include the details of a class assignment and whether you plan to publish the profile and where. To get a response from the scientist, make explicit the amount of time you expect the interview to take. Indicate your availability and your willingness to send some of the questions ahead of time should they be willing to meet with you. Once they've said yes, you might ask whether you can record the interview ahead of time so they don't feel put on the spot. Give your reasons for wanting to record, which generally include ensuring accuracy in the final report and giving yourself the freedom to listen during the interview.

The Interview Process

The interview is the heart of profile writing, but it's different from interviewing for a news story or academic paper. You're not just gathering facts or quotes—or following a scripted research interview protocol as you would to generate publishable research—you're trying to understand who someone is and how they move through the world. The following suggestions are informed by an interview workshop given by the Stanford Storytelling Project (Hartman, 2022):

Before the Interview

- Interview yourself about your motives. Spend some time reflecting in writing on why you've selected your subject and what you hope to learn about how they pursue their work.
- Prepare open-ended questions that invite storytelling rather than yes/no answers.
- Plan to ask about specific moments, not just general attitudes.
- Think about what you want to observe as well as what you want to hear; that is, don't just think about the content questions you don't have answers to, but also—and perhaps more importantly—about the questions you have regarding the scientist's character or process.

During the Interview

- Ask again whether you have their permission to record. You can say that it will help confirm the accuracy of quotations. If they do not want to be recorded, you will need to be prepared to take very good notes.
- Listen with compassion for plot points—moments of change, decision, or discovery.
- Ask for specific details that help you visualize scenes.
- Don't be afraid to ask for clarification or elaboration.
- Pay attention to your subject's physical presence, mannerisms, voice, and environment.

Sample Interview Questions

- "Walk me through what happened" (to elicit a chronological narrative)
- "Where were you when?" (to ground abstract events in specific places)
- "What were you thinking [or feeling] at that moment?" (to access interior experience)
- "Is that like ...?" or "How would you explain that to an advanced undergraduate?" (to elicit analogies and metaphors)
- "How did that change you?" (to understand growth or transformation)
- "Do you think we left out anything the readers should know?" or "Who else could tell me more about what we discussed today?" (to conclude the interview)

Observation and Scene-Setting

If your subject is local and they consent, go to them for the interview. You want to spend time with your subject in their natural environment. Watch them work, interact with colleagues, or navigate their daily routines. These observations provide the concrete details that bring profiles to life. Notice what's in their workspace, how they explain concepts to students or labmates, and the expressions on their faces as they think.

If you cannot observe your subject directly, ask them to describe important scenes in detail. For example, you might ask them: "Tell me what the laboratory looked like that day." "What did the hospital room smell like?" "Who else was there, and what were they doing?" These questions help you reconstruct scenes even when you weren't present.

Reflecting on and Interpreting the Interview

It's easy to treat interviews as pure, unfiltered access to someone's experiences—but this assumption can lead us astray. Every interview represents two interconnected processes: an act of memory and a conversational exchange between two people.

Memory itself is imperfect. Even well-intentioned interviewees regularly misremember dates, mix up names, or blend separate incidents into a single narrative. Sometimes the inaccuracies are deliberate—people may embellish details to enhance their reputation, protect others, or serve personal motivations.

Meanwhile, interviewers bring their own influence to every conversation. Despite claims of neutrality, each interviewer shapes the dialogue through their choice of questions, the topics they pursue or abandon, and the assumptions they bring to the exchange. The result? The same person might tell markedly different versions of identical events depending on who conducts the interview.

Understanding these dynamics doesn't diminish the value of interviews —it simply reminds us to approach them as complex social interactions rather than straightforward data collection. You will need to reflect on the credibility of, and potential biases in, the interview material you collect.

The following are some questions you might answer as you review your interview transcript, adapted from Linda Shopes's "Making Sense of Oral History" (2002):

Who is the interviewer?

- How did your background and interests affect the interview?
- How did your questions shape the story told?
- What effect might your social identity and position have on the interview?
- How might your purpose have shaped the content, perspective, and tone of the interview?

Who is the interviewee?

- What is the interviewee's relationship to the events under discussion?
- What stake might the interviewee have in presenting a particular version of events?
- What effect might the interviewee's social identity and position have on the interview?
- How does the interviewee present himself or herself in the interview?

What is being said in the interview?

- What does this tell us about the way the person being interviewed thinks about his/her/their experience?
- What does the interviewee sidestep?
- What does the interviewee especially warm to or discuss with interest and what might this tell us?
- How does the interviewee's account compare with other sources and interviews?

As you draft your profile, you will want to think about how much of the answers to these questions you should make explicit for your readers.

Structuring Your Profile

Most profiles follow a flexible structure that alternates between "showing" (scenes where we see the subject in action) and "telling" (background information, analysis, or reflection). This creates rhythm and helps readers process both story and information.

Common Profile Structures, Which Can Be Used Alone or in Combination

Chronology: Follow your subject through a specific period or process from beginning to end. This works well for profiles focused on a particular discovery, treatment, or project. It will be most interesting if it acknowledges setbacks or challenges and the ways the scientist worked to overcome them.

Theme: Organize around key aspects of your subject's character or work, using scenes to illustrate each theme. This works well for profiles of established figures with complex careers.

Inventory: Take stock of your subject's workplace, habits, or relevant papers. A list that reveals a pattern of some kind renders a coherent character.

Turning Point: Begin with a crucial moment, then flash back to show how your subject reached that point, and finally move forward to show the consequences. This creates natural dramatic tension.

As you develop your story arc, consider where you will weave in the science, which cannot be left out. Often, it will come in as you discuss your subject's motives (what was not yet known?), the controversy they responded to (what did the various camps of scientists believe?), or their contribution (how did they advance knowledge?). Work to explain at least some of this in concise, everyday language.

Developing the Character of Your Profile Subject

Strong profiles require more than just good reporting—they require artful writing that brings your subject to life on the page. This means paying attention to, and rendering, character in words.

As you draft your profile, think about how you can add the following:

Concrete details: Instead of merely saying someone is "passionate about their work," show them staying late in the lab, talking excitedly about data, or pacing while they explain a concept to a group of students. If you're profiling a vaccine skeptic who's converted to a vaccine activist (or vice versa), for example, you will want to document in some detail the particular scientific paper, conversation with a doctor or patient, or family experience that changed their mind and behavior. In Carolyn Johnson's 2025 *Washington Post* profile of a patient receiving experimental cancer treatment, we learn that he is a gardener, golfer, lawyer, and father. When we meet Richard Schlueter, he is planting cucumbers and squash in a community garden.

Distinctive voice: Capture how your subject actually speaks, including their particular vocabulary, rhythm, pace, volume, and manner of expression. If you've come to a press conference to listen to your subject communicate their latest findings, you may want to capture the emotion—or lack thereof, depending on the choices made—that is audible: the tremor or pitch of the voice. Schlueter, in his reflection on his cancer, mixes everyday language with the technical language of medicine. Referring to his legal training, he says, "I'm a fact-finder, and I spend a lot of time researching and trying to find a path to remission or a potential cure. And it's a pretty bleak outlook" (Johnson, 2025).

Physical presence: Help readers visualize your subject through specific, relevant details about their appearance and movements, especially as they interact with their work. If you're describing the way a psychologist interacts with young patients, for example, you may want to describe the care they take with their hands and posture to communicate calm or, conversely, the energy they bring to their steps or gestures to keep a child's attention. In Johnson's profile, we learn that Schlueter's "pain in his hip makes it hard to get out of a chair" (Johnson, 2025).

Internal life: Use your subject's own words and anecdotes—little stories within the larger story—to reveal their thoughts, motivations, and emotional responses to failures and successes. Readers will want to know what a scientist was thinking in the face of an experimental setback, funding cut, or community censure. How did they persist? In Schleuter's case, we learn that he was recently well enough to run, bike, and swim, and that he continues to care for his health because he hopes the experimental treatment will advance medical science and help others plagued by cancer (Johnson, 2025).

Complexity: Highlight how your subject is multidimensional and capable of change. For example, an ecologist, an expert on trees, may also be a mother, and an intriguing profile of this person might explore how both facets of their identity contribute to their thinking about the future. In the Johnson profile, we get to know Schlueter, but we also come to understand the complex bureaucratic context that impacts access to experimental cancer treatment. What's changing in this piece is government funding for science (Johnson, 2025).

Revision and Fact-Checking

Profiles require especially careful revision because they blend factual information with narrative storytelling. Check all scientific information for accuracy, verify quotes with your subject when possible, and ensure that your characterizations are fair and evidence-based.

Be wary of the following pitfalls, too:

Focusing too much on biography or chronology: Unless someone's personal history is directly relevant to their scientific work, don't get bogged down in childhood stories or comprehensive life narratives.

Trim away any background that doesn't directly serve the narrow slice of a scientist's career that you're addressing. Profiles are snapshots, not feature-length films.

Avoiding the science: Profiles should illuminate scientific concepts, not just personality. Make sure readers understand not only what your subject actually does and why it matters, but also the technical background information they need to make sense of the story.

Creating false drama: Don't exaggerate conflicts or oversimplify complex issues just to create narrative tension. Don't reinforce the myth of an individual scientific genius. The real drama in science often lies in subtle moments of new understanding and quiet victories won through collaboration.

Losing focus: Remember that your profile should have a clear point or theme. What do you want readers to understand about science or technology through your subject's story?

Blindness to your own bias: As you assess your draft, be aware of the assumptions, prior knowledge, and biases you have brought to your representation. How has your connection to the subject informed your profile? Your readers will trust you more as a communicator if you explain the background knowledge they need to understand the science and if your representation is balanced—neither an uncritical celebration of someone nor an indictment of their character.

Once you're content with the content and organization of your profile, share your complete draft with a new reader, someone who knows nothing about your subject or topic. After they've read your profile, ask them: What's your impression of this person? What scientific understanding did you gain? Their answers should help you determine whether you're achieving your purpose. If not, keep working!

Lastly, consider sharing your profile with your subject before publication. While they shouldn't have editorial control over your work, they can catch factual errors and help you understand whether you've missed important context or misrepresented their views.

TAKEAWAYS

- Profiles humanize science by focusing on individual stories that explain the process of science or demonstrate its impact.
- Strong profiles require three essential elements: vivid setting, compelling character, and meaningful plot.
- Effective interviewing involves asking for specific scenes and details, not just general information.
- The best profiles alternate between showing (scenes in which the making of science falters or advances) and telling (background information and analysis of the action).
- Profiles carry ethical responsibilities to represent subjects fairly and accurately.

FURTHER RESOURCES

Association of Healthcare Journalists. (2025). Statement of principles of the Association of Healthcare Journalists. https://healthjournalism.org/about/principles-and-policies/statement-of-principles-of-the-association-of-health-care-journalists/

Gladwell, M. (2007). The limits of profiles. In M. Kramer & W. Call (Eds.), *Telling true stories: A nonfiction writers' guide* (pp. 73–74). Plume.

Hart, J. (2021). *Storycraft: The complete guide to writing narrative nonfiction.* University of Chicago Press.

Wilkerson, I. (2007). Playing fair with subjects. In M. Kramer & W. Call (Eds.), *Telling true stories: A nonfiction writers' guide* (pp. 172–176). Plume.

REFERENCES

Banaszynski, J. (2007). Profiles. In M. Kramer & W. Call (Eds.), *Telling true stories: A nonfiction writers' guide* (pp. 66–69). Plume.

Bowden, M. (2012, August). The measured man. *The Atlantic.* https://www.theatlantic.com/magazine/archive/2012/07/the-measured-man/309018/

Carey, B. (2017, May 15). Brenda Milner, eminent brain scientist, is 'still nosy' at 98. *The New York Times.* https://www.nytimes.com/2017/05/15/science/brenda-milner-brain-cognitive-neuroscience.html

Fitzpatrick, Ciara C. (2025, September 2). Interview with Dr. Caleb Scharf. *Interview conducted by Fred Van Wert on March 28, 2024.* National Aeronautics and Space Administration. https://www.nasa.gov/general/interview-with-dr-caleb-scharf/

Gawande, A. (2007, April 23). The way we age now. *The New Yorker*. https://www.newyorker.com/magazine/2007/04/30/the-way-we-age-now

Hartman, C. (2022, April 18). *Interviewing for a story workshop*. Program in Writing and Rhetoric, Stanford University.

Johnson, C. Y. (2025, August 18). His custom cancer therapy is in an NIH freezer. He may not get it in time. *The Washington Post*. https://www.washingtonpost.com/science/2025/06/18/nih-cancer-therapy-delay-staff-cuts/

Max, D. T. (2017, December 11). Jim Simons, the numbers king. *The New Yorker*. https://www.newyorker.com/magazine/2017/12/18/jim-simons-the-numbers-king

Remnick, D. (2000). Introduction. In D. Remnick (Ed.), *Life stories: Profiles from The New Yorker* (pp. ix–xii). Random House.

Shopes, L. (2002). Making sense of oral history. *History matters: The U.S. survey course on the web*, https://historymatters.gmu.edu/mse/oral/.

Yasmin, S. (2018). *The impatient Dr. Lange: One man's fight to end the global HIV epidemic*. W. W. Norton.

CHAPTER 8

Podcasts

Communicating Science for the Ear

WHAT YOU'LL LEARN IN THIS CHAPTER

- A brief history and genre map of science podcasting
- Key rhetorical principles of effective science audio for public audiences
- How technology and AI are reshaping the medium of podcasting
- A step-by-step process for making your own podcast
- Hands-on activities to reinforce practice with steps in the podcasting process

INTRODUCTION

In today's attention-scarce media landscape, a single well-produced podcast episode can often reach more people and communicate scientific ideas more effectively than an academic article. Done well, audio communication unfolds complex research through voice, pacing, narrative, and metaphor, allowing listeners to follow patterns of reasoning over time and draw meaningful insights without needing specialized training. For these reasons, podcasting has become a cornerstone of contemporary science communication.

But not every science podcast enlightens. Poorly designed episodes that use heavy jargon, lack clear structure, or make oversimplified claims can

DOI: 10.1201/9781003520887-10

just as easily confuse or mislead public audiences. The conversational tone and affective intimacy of the audio medium can amplify misunderstanding just as much as they support public insight. So how do scientists use the medium to communicate science effectively?

As you will see in this chapter, effective science podcasting depends on a set of core principles that attend to audience, structure, evidence, and the affordances of audio in order to design podcasts that inform, engage, and responsibly communicate scientific knowledge. In the remainder of this chapter, we move from history to principles, to process, and finally to hands-on practice. By the end, you'll have a practical toolkit for planning, producing, and sharing your own science podcast—one that translates research into stories that listeners can follow, remember, and trust.

GENRE IN CONTEXT: HISTORY AND EXAMPLES

Before smartphones, science reached public ears mainly through radio—programs like the BBC's *Reith Lectures* (1948–present), NPR's *Talk of the Nation: Science Friday* (1991–present), and *Quirks and Quarks* from CBC Radio (1975–present). These programs followed *broadcast logic*: fixed schedules, one-to-many delivery, and limited replay. All that changed in 2004, when the fusion of "audio" and "RSS feed" gave birth to podcasting—an on-demand, mobile medium. Two decades after the birth of podcasting, over half a billion people listen to podcasts weekly.

Freed from broadcast constraints, programs like *The Naked Scientists* (Cambridge University, 2005–present), *Skeptics' Guide to the Universe* (2005–present), and later *Radiolab* (2002–present) experimented with length, structure, and sound design, building global niche audiences who otherwise may not have been interested in or able to access science in print. In short, podcasting has helped democratize science storytelling, putting narrative control and audience choice directly into the hands of creators and listeners.

For science communicators, the ear offers a direct path to the imagination, turning commutes or chores into moving classrooms. As an aural medium, podcasts are both intimate and flexible in structure. Yet the very things that make podcasts so appealing for listeners—narrative freedom, conversational tone, and low technical barriers—demand a skill set different from writing peer-reviewed prose: How do you turn data tables into narrative scenes? Or explain CRISPR to someone chopping vegetables? This chapter walks you through some of the more popular types of science podcasting and provides you with a series of principles and processes, including activities, that give you a roadmap for creating or improving a science podcast that meets audiences where they are and draws them toward where science is going.

When you scan the science section of Apple Podcasts or Spotify, you'll notice that most shows fall into a few recurring formats. Many podcasts mix elements from different genres, but each genre has a core learning function and a distinct emotional payoff—two factors that heavily influence whether an audience sticks around for the next episode. Table 8.1 outlines the six most common genres of science podcasting and their educational and emotional goals, along with an explanation of why they work well and sample podcasts you might explore for further inspiration. As Table 8.1 demonstrates, different science podcast genres meet different audience demographics and needs: an *explainer* format works for someone with 2 minutes on a lunch break; a *narrative documentary* might engage a weekend listener on a long drive. As you explore genres, think of them as design choices—not just containers for information, but structures that shape how audiences process, feel, and remember science.

TABLE 8.1 Core Genres in Contemporary Science Podcasting

Genre	Learning Goal	Emotional Goal	Why It Works	Example Podcasts
Explainer/ Bite-Size	Deliver a clear, concise answer to one question	Spark curiosity and satisfaction	Short, focused episodes fit into busy schedules and lower cognitive load, which improves recall for non-experts.	*60-Second Science* (Scientific American), *TED-Ed Audio*
News Round-Up	Summarize and contextualize recent research	Build trust and a sense of being informed	Naming sources and experts signals credibility, while added context keeps audiences current .	*Nature Podcast*, *Science Weekly* (The Guardian)
Interview and Panel	Share expert insights through conversation	Create connection through personal stories	Hearing scientists in dialogue humanizes research and increases emotional engagement, which boosts memory retention.	*Ologies with Alie Ward*, *Huberman Lab*, *Science Vs*

(*Continued*)

TABLE 8.1 (*Continued*) Core Genres in Contemporary Science Podcasting

Genre	Learning Goal	Emotional Goal	Why It Works	Example Podcasts
Lab Notebook/ Process Diary	Show how science is done in the field or lab	Evoke authenticity and discovery	Revealing the process behind results builds transparency and helps audiences understand why uncertainty is part of science.	*The Naked Scientists*, *Antarctic Stories*
Kids and Family	Teach age-appropriate science concepts	Inspire joy, play, and shared learning	Simplifies complex ideas without "dumbing down," using play and curiosity to make learning a shared, memorable experience.	*Brains On!*, *Wow in the World*
Narrative Documentary	Explain complex topics through story arcs	Create wonder and emotional investment	Story-driven arcs keep listeners engaged through technical content and increase empathy for the people and ideas involved.	*Radiolab*, *Hidden Brain*

CORE PRINCIPLES AND RHETORICAL STRATEGIES FOR COMMUNICATING WITH SOUND

A great science podcast isn't just about what you say; it's about how you say it. Audio is ephemeral: once words are spoken, they vanish, and most listeners won't rewind, nor are transcripts always available. That means every sentence, sound cue, and transition must work intentionally to keep listeners oriented, help them understand complex ideas, and earn their trust—what McHugh (2022) calls the "invisible art" of learning to use voice and soundscapes effectively to build connection and authenticity with listeners. This section introduces five rhetorical strategies that help science communicators meet those challenges: story, metaphor, sound, signposting, and transparency.

Story

As we've already previewed in Chapter 2, *Rhetorical Tools*, storytelling is a powerful tool for engaging public audiences with complex scientific topics, and it has a special place in podcasting As Dahlstrom (2014) explains, story provides a powerful cognitive and emotional scaffold for learning science. By establishing the stakes, mood, or human perspective first, story makes scientific ideas meaningful. A strong story also scaffolds comprehension by giving audiences something to hold onto—whether a character, a question, or a vividly rendered world—so that technical concepts arrive with context rather than in a vacuum.

Beyond cognitive scaffolding, story anchors us to people, places, and the broader moral stakes at play in scientific progress, establishing a rich sense of place for listeners. In Overheard at National Geographic's episode "The Canary of the Sea," listeners are invited into vivid field audio and a description of the fog-shrouded Arctic where beluga researchers work (Briggs, 2020). This immersive introduction establishes the mood and context of the frigid Arctic seas before the episode shifts to explain how underwater noise interferes with beluga communication. The setting becomes the conceptual framework for understanding the later scientific details.

This narrative approach in podcasting is effective because it taps into an ancient oral storytelling tradition; we are better primed to listen when we know who and what the science is about, where it happens, and why it matters. Consider the Radiolab series "Gonads: Dutee" (Webster, 2021). Instead of leaden technical discussions of endocrinology, the episode begins with Dutee Chand's childhood memories of running barefoot, where "tiny pebbles in the road would get stuck in [her] feet." By the time the episode fast-forwards to her record-breaking career and the subsequent controversy over her biological test results, the audience is personally invested in her biology. The technical complexities of endocrinology are no longer abstract; they are high-stakes hurdles for a character we've come to know and care about.

Indeed, effective science podcasting often uses storytelling to raise compelling questions about the moral and ethical stakes of science—explaining why science matters. For "Gonads," it's partly about the moral dilemma of gender verification testing in elite women's athletics and its high-stakes forathletes like Dutee (Webster, 2021). In Radiolab's "Antibodies Part 1: CRISPR," it's about the way gene-editing technologies like CRISPR "may someday let us change the course of human evolution" (Abumrad & Krulwich, 2015). Finally, "Canary in the Sea" presents an ecological moral dilemma about the conflict between human industrial "progress" (shipping, oil exploration, military sonar) and the fundamental biological necessity of animal communication these technologies imperil.

By anchoring listeners in these moral dilemmas, science podcasters create powerful narrative hooks that sustain listeners' curiosity through the more technical explanations that follow.

As we outlined in the introduction to this book, and as you will see demonstrated in other chapters throughout this book, story is not ornamentation but the architecture of effective science communication. By framing science within vivid environments, human journeys, and urgent ethical questions, story provides podcast episodes with a structure that helps audiences invest emotionally, absorb complexity, retain key insights, and, importantly, understand why science matters.

Metaphor

A metaphor is a comparison that maps the features of something familiar onto something unfamiliar, allowing listeners to understand new concepts through what they already know. Metaphors are especially important in science communication, where technical jargon and abstract concepts or processes—gene editing, chemical signaling, climate feedback loops—can overwhelm or confuse listeners (Lindenfeld et al., 2025). Metaphors help people see the science in action by turning abstract concepts into familiar mental pictures that make science more understandable and memorable. For example, in the Radiolab episode "Antibodies Part 1: CRISPR," scientist Jennifer Doudna explains how CRISPR acts as a text editor that allows scientists to change the genetic code of animals, plants, and microorganisms (Abumrad & Krulwich, 2015). Doudna likens the complex gene-editing technology to a pair of "smart" molecular scissors used to cut the DNA. By comparing a complex technology to familiar objects like scissors, the metaphor gives listeners an instant visual foothold, which the episode can then refine with accurate scientific terms.

In audio storytelling, where no images are available, effective metaphors are a staple of the science communicator's toolkit, acting as cognitive scaffolds that keep listeners oriented, engaged, and able to follow progressively more complex explanations.

Sound

It should come as no surprise that in a sonic medium, sound is one of the most powerful rhetorical tools available to science podcasters. In audio storytelling, sound often does the work that visuals or diagrams do in other media: it anchors ideas in sensory experience, establishes mood, and invites audiences to inhabit scientific spaces they may never physically enter (McHugh, 2022).

Like story, sound plays a crucial role in helping to *set the scene* by creating engaging *soundscapes*—a layered combination of ambient noise, music, and voice—that transport listeners into specific scientific contexts. If the story is the shape of the episode's narrative arc and its main characters, then sound is the texture and feeling of the episode as we move through the story, helping listeners imagine the people and places where science happens. This is key for both listener comprehension and engagement. In a *Radiolab* episode titled "Playing God," producers used soundscapes to recreate the atmosphere of a hospital in crisis post-Hurricane Katrina (Abumrad & Krulwich, 2016). Listeners experience a shift from steady hospital hums to the rhythmic chirping of failing ventilator batteries, layered with eerie music and interviews with doctors recounting the "pitch blackness" and "absolute emergency" of the crisis. This sensory immersion places listeners in a tangible reality before the narrative introduces complex bioethical concepts, such as "deliberative democracy," explored later in the episode. By forcing the audience to first "hear" the environment of a failing hospital, the producers use audio as a primary tool for framing the moral weight of practicing medicine during resource scarcity. In this way, sound, like story, is also used to *set the emotional tone* of an episode and establish the moral weight of science and its impact on individuals, communities, and our broader world.

Finally, when field recordings make the science audible, sound functions as *evidence*. This is crucial for the listening ear. In a medium where listeners can't "see" scientific processes or look at data and charts, sound clips from the field place listeners in the moment and authenticate the scientific methods and evidence discussed in the episode. Consider, for example, how in the episode "The Canary of the Sea" from National Geographic's *Overheard*, listeners hear distinct beluga vocalizations in the wild before the scientific explanation begins (Briggs, 2020). When marine biologist Valeria Vergara then explains how increasing interference from ship noise masks whales' calls and their ability to communicate with each other, the contrast between natural sound and human culture gives listeners experiential proof of the issue.

Consider how sound is used in the *Radiolab* episode "Sound as Touch" as a form of *embodied evidence* to bridge the gap between abstract physics and physical sensation (Abumrad & Krulwich, 2007). The episode tells the story of a woman who is profoundly deaf but can "hear" her husband's voice by resting her hand on his cheek or chest, feeling the specific vibrations of his vocal cords. To help the audience understand this scientific phenomenon, the producers employ an intimate, close-mic sound design to remove all background reverb as a tool for sensory translation, mimicking the skin-to-skin contact required for the woman to perceive her husband's words and making the husband's voice feel as though it is whispering directly into the listener's ear. By making the audio feel physically "close," the podcast invites the listener to experience the same sense of tactile intimacy, transforming a mechanical biological process into a palpable, human connection.

Like story and metaphor, sound in science podcasts is not there for background decoration. Rather, it is a fundamental storytelling tool that transports listeners into scientific environments, deepens listener engagement, and illustrates complex scientific concepts to enhance listener understanding.

Signposts

Because audio disappears the moment it is spoken, listeners benefit from clear signposting—verbal cues that structure information and guide comprehension. Here are three main ways science communicators build cohesive structure and aurally "signpost" content for listeners:

Roadmaps are a key strategy for orienting listeners of science podcasts. In many episodes of *Science Vs*, host Wendy Zukerman utilizes signposting to preview the structure of her episode. In "Fasting Diets: What's the Skinny?" Zukerman (2019) signposts the episode's structure by stating, "Today we're going to look at three main reasons that people are doing this. *First*, can it [fasting] help you lose weight? *Then*, it is better for your metabolic health? *And finally*, could it help you live longer?" This simple roadmap gives listeners mental "folders" to accurately anticipate and organize new information as it arrives.

Repetition is another essential tool for managing listeners' cognitive load. Science podcasters often use this technique to strategically reintroduce a previously defined concept to ensure the listener maintains the necessary context to process new information as the episode progresses. As mentioned earlier, because audio is a linear medium, listeners cannot "flip back" to double-check a definition; repetition works as a mental anchor to stabilize listener understanding. We can see this at play in the *Ologies* podcast episode "Virology (Coronaviruses)," where host Allie Ward (2020) introduces the mechanism of viral replication as a "hijacking" of a cell's genetic instructions. She later updates this definition, using the more vivid analogy of a "photocopier heist," to explain how the virus tricks the cell into making copies of the virus instead of itself. By re-explaining the biological mechanism through a different lens later in the episode, Ward reinforces the core science for the audience, ensuring that the complexity of the viral "hijack" remains clear as the discussion shifts toward treatment and prevention.

Moving from *simple to complex language* also supports clear communication in podcasting. In discussing how a virus like COVID-19 spreads, for example, introducing plain-language descriptions helps listeners build an intuitive understanding of the topic before encountering unfamiliar terminology. This scaffolding technique is a hallmark of the podcast *Science Vs*, where host Wendy Zukerman (2020) describes the coronavirus as a "spiky little ball" that uses its surface proteins like "keys to open a lock" on human cells—notice how this simple-to-complex language sequencing also relies on the use of metaphor, a principle we discussed earlier. Only after establishing this mechanical metaphor does the episode introduce the technical name for the lock: the ACE2 receptor. By sequencing language this way, the podcast ensures the listener understands the biological mechanism before being asked to memorize the nomenclature.

Transparency

One final principle of effective science podcasting we want to emphasize is transparency. Now more than ever, when public trust in institutions and science is waning, transparency helps scientists build back trust not by asserting authority, but by showing how knowledge is constructed and conceding where knowledge is limited. Transparency about evidence, process, and uncertainty helps listeners better evaluate claims and understand science as a dynamic, iterative practice. It is a cornerstone of the scientific ethos. Here are three techniques you can use to build transparency into your podcasting practice:

Source Transparency

Being clear about where you get your information from is crucial to building credibility in science communication. For example, in the *Science Vs* podcast "Fasting Diets: What's the Skinny?" (2019), Zukerman brings on experts—Dr. Krista Varaday, who has led trials on alternate-day fasting, and Dr. Courtney Peterson, who specializes in time-restricted feeding—cites specific clinical trials, explains sample sizes, notes dropout rates, and clarifies the limits of rodent studies,. These practices make scientific evidence both visible and legible and model how to interpret scientific claims responsibly.

Process Transparency

Another strategy to consider for strengthening your credibility is discussing your methods and scientific process. In *Overheard* at National Geographic's "The Canary of the Sea" (2020), host Amy Briggs explains *how* the audience is hearing real hydrophone audio of field recordings of beluga sounds captured in the field by scientist Valeria Vergara. These production details make the sounds—and therefore the science—feel trustworthy.

Acknowledging Uncertainty

Uncertainty is part of science and acknowledging and explaining this to listeners is equally important in building credibility. Consider, for example, Zukerman's episode on "Coronavirus" (2020), which tells listeners, "There's still so much we don't know. We don't know how easily this [corona]virus spreads, we don't know how deadly it is, and we don't know how many people have actually had it." By naming gaps in knowledge and interviewing epidemiologists about unstable early data, the episode demonstrates epistemic humility that strengthens, rather than weakens, audience trust.

TOOLS, TECHNOLOGY, AND THE ROLE OF AI IN PODCASTING

Before exploring tools and technologies, it is essential to clarify the role of technology, and especially AI, in podcast production. AI can speed up many technical tasks in podcast production (transcription, noise reduction, first-pass editing), but it cannot replace human judgment, curiosity, or ethical responsibility. In science podcasting especially, accuracy, transparency, and narrative intent must remain in the hands of human storytellers who think critically and communicate with care. Overreliance on AI can flatten personality, introduce error, and erode your voice as a storyteller. Use AI as an optional assistant—not as an author and not as a requirement. You are still responsible for verifying every claim, crediting every source, and making interpretive choices with intention.

Essential Tools: What Beginner Podcasters Actually Need

For beginning podcasters, the essential tools are more accessible than you might think. You can create a strong, well-produced episode using devices you already own, such as a smartphone and free editing software. A basic podcast setup includes the following three components: something to record with, something to edit with, and a quiet space to capture clean audio. From there, you can scale up. Campus libraries often lend USB microphones and portable field recorders, and many institutions provide access to entry-level audio studios. As you advance, you may choose to work with more sophisticated software for multitrack editing, noise reduction, mixing, and mastering. What matters most is learning how to use the tools you have with intention and care, so your storytelling stays clear, credible, and compelling. Table 8.2 helps you navigate these choices, distinguishing which tools are necessary for producing a basic episode and which ones you can add later as your skills, resources, and ambitions grow.

How AI Is Transforming Podcasting

While human editorial judgment remains the primary driver of high-quality audio content, artificial intelligence (AI) has become an integrated tool in

TABLE 8.2 Beginners Guide to Podcasting Tech

Category	**Must-Haves (Beginner-Friendly)**	**Nice-to-Haves (Advanced or Optional)**
Recording Device	Smartphone voice memo app; basic laptop mic	USB microphone, campus-loan field recorders
Recording Environment	Quiet room, closet "booth," blankets or clothes to reduce echo	Acoustic foam panels, portable isolation shields
Editing Software	Free tools: Audacity, GarageBand	Adobe Audition, Reaper, Hindenburg Pro, Descript
Headphones	Any wired headphones (avoid Bluetooth lag)	Closed-back studio headphones
Field audio	Smartphone for short ambient clips	Dedicated field recorders; external lavalier mics
Scripting and planning	Google Docs, Word, Notes app	Two-column script templates; AI-assisted drafting or organization tools
AI tools (optional)	Free transcription tools (e.g., Otter's free tier), basic noise reduction	Descript Studio Sound, AI script polishers, AI idea organization

many podcast software applications for managing the technical and organizational complexities of podcasting. In current practice, AI is utilized to automate repetitive production tasks such as generating episode outlines, streamlining audio editing through text-based transcripts, and enhancing sound quality via automated noise reduction. These applications function as a supplement to, rather than a replacement for, the foundational skills of a producer.

Still, the integration of AI in science communication necessitates a rigorous ethical framework to ensure public trust and scientific accuracy. As AI models are prone to "hallucinations," or the generation of misinformation, producers must maintain transparency regarding the use of synthetic tools and prioritize manual fact-checking. Furthermore, the use of generative AI must be balanced against intellectual property rights and the responsibility to represent scientific data without bias or oversimplification. Used thoughtfully and responsibly, AI can cut production time significantly, freeing you to shape stronger stories, refine emotional arcs, and guide listener curiosity with more nuance.

Science podcasters must ensure accuracy, provide transparent citations, and guard against bias and misrepresentation. We advise students to treat AI usage as an option, not a mandate. Furthermore, if students find themselves using AI in the scripting or editing process, they should do so conservatively, with an eye toward significant human intervention and revision to align with professional standards and audience values. They should also be prepared to disclose and align usage with the rules of the publication or instructor guidelines. At the end of the day, ethical stewardship of your podcast and its content remain your responsibility.

A PROCESS FOR CREATING YOUR OWN SCIENCE PODCAST

You've explored the genres, learned the rhetorical strategies, and seen how AI can streamline the technical work. Now it's time to put it all together. This six-step process walks you from idea to publication, showing how to plan, record, and release a science podcast that's accurate, engaging, and ethically sound. Think of it as a production map: each step builds on the last, combining creative storytelling with clear communication and trustworthy sourcing. Whether your goal is a 5-minute explainer or a 30-minute deep-dive, these steps will help you structure the science and the listening experience so your audience leaves informed, inspired, and ready to press "play" again.

To make the process concrete, we'll follow a hypothetical example all the way through: a 10–15-minute explainer episode answering one deceptively simple question—*Why are flamingos pink?* At each step, you'll see how to weave in core rhetorical strategies such as *story scaffolding, metaphor, sonic evidence, inclusive framing,* and *strategic repetition* to successfully communicate our research to the public.

Clarify Purpose and Audience

Every great science episode starts with a question worth answering. This core question guides your research, shapes your narrative arc, and informs your choice of sounds. It's the compass for every creative and editorial decision you'll make. For example, let's imagine you're in an entry level biology class and you are currently studying the relationship between an organism's phenotype (physical appearance) and its environment. Your class takes a field trip to the local zoo where you notice a group of flamingos. The question suddenly strikes you: *Why are flamingos pink?* This is an example of what we mean by starting from a question worth answering: "Why are flamingos pink?" is a specific, answerable question that opens the door to bigger topics like diet, biology, and ecosystems.

Once you decide on your topic, the next and arguably most important thing to consider is your audience. To whom are you speaking? Your audience determines many choices for your podcast: pacing, language choice, and how you frame the conversation. Continuing with our flamingo example, let's say you are interested in explaining the concept to middle schoolers. Now that you have your core question, let's try defining your episode goal in a single, clear sentence. This sentence should include the goal of your question and your audience. This would look something like: Episode 01: *"To explain how flamingos get their pink color to curious middle schoolers."*

Choose Format and Genre

Once you have your big idea and you've established the purpose and audience of your episode, the next challenge is figuring out *how* to tell the story so that your audience cares from the very first moment. Whether it's a personal journey, a scene-rich documentary, or a tight news briefing, your genre determines elements of the episode's format such as pacing, tone, and where you place your big reveals. Match the episode's format to your goals. Quick tips? Use an *explainer* format. Deep dives? Try *narrative documentary.* A clear frame helps you decide how much science to frontload and how much to hold back for later in the episode. With that frame in mind, you can now gather the raw material—the facts, voices, and sounds—that will bring your story to life. Staying with our flamingo episode, remember you're speaking to middle schoolers, so you need to be informative but concise. Therefore, you pick a short *explainer* format. You think about how best to use *sonic evidence* to help create the beginning of the *story* and consider opening with a sound-rich scene from field audio in a wildlife refuge: squawks and splashes to set the scene and, if possible, integrating clips from an interview with a zoologist to serve as a narrative guide, like Ms. Frizzle in *The Magic School Bus.*

Research and Script

Once you've established your episode's purpose, audience, and format, the next challenge is scripting the episode's arc. This involves research (both reading and field work) and scripting. Begin with accessible, credible sources like expert interviews, key research papers or reference material, and relevant sound cues from field audio. You will want a blend of field audio as a sound bed to bring authenticity and credibility to the episode, and, if possible, other voices to increase engagement and heighten persuasive appeal.

Once you've gathered these foundational materials, your next step is to dig deeper into your sources and shape them into the intellectual backbone of the episode. Review primary sources such as original studies and research, pre-interview guests, and collect key anecdotes and data that help you clearly explain your big idea.

Unless you are doing an interview format, we strongly recommend scripting. Scripting helps you create structure. Scripting isn't just deciding *what* to say; it's designing *how* the listener will experience the information moment-by-moment. Once your script is in place, you're ready to capture audio that matches its tone and clarity.

Finally, remember that podcast listeners can't re-read a sentence if they get lost. As you write your script, here are a few tips we recommend to help readers follow along more easily:

Write Conversationally for the Ear

In practice, this means using shorter sentences, active verbs, and auditory signposts ("here's the twist," "think of it this way," "first let's look at") that help listeners follow along. Think of how you'd explain the idea to a curious friend—not dumbing it down, but stripping away academic clutter so the meaning arrives cleanly. Always read aloud to test pacing and clarity.

Use a Two-Column Script to Outline Narration and Sound Cues Together

Using a two-column script helps you keep your narration aligned with your soundscapes. Plan where to deploy rhetorical strategies—such as metaphors, story scaffolds, or sonic evidence—so the science lands with clarity and impact. For example, you might place a metaphor right before a complex explanation to anchor the listener, or consider bringing in a piece of ambient sound just as you shift scenes to signal movement and maintain engagement.

Introduce Your Sources in a Clear and Easy-to-Follow Manner

In audio, oral citations should be brief, clear, and woven seamlessly into the narrative, including source attribution information that signals the author or source of the information and the date. For example, try something like, "A 2023 study in *Nature* found…" or "As biologist Maria Torres told me when we spoke last week…". The goal is to give listeners enough information to understand the source's credibility without interrupting the rhythm of the episode. When you summarize a study, foreground the insight that matters for your story, not the methodological details. For even more credibility, include a list of sources in the episode's *Show Notes* page.

Here's how these tips would work with our hypothetical flamingo episode. You might begin by drawing on ornithology papers about carotenoid pigments, interviewing a zookeeper, and reviewing accessible science explainers. For instance, you might say, "According to a 2021 study in the journal *Avian Biology*, wild flamingos get more than 80 percent of their pink pigment from tiny algae called diatoms…" and then move directly into your explanation of how that pigment accumulates. You can pair this with a field interview clip to provide a vivid, human anchor that bridges the science to everyday language: "As zookeeper Daniel Ruiz told me when we spoke last month, 'They're basically eating color.'" You then build your story scaffolding around Ruiz's feeding observations, introduce a "tiny paint drops" metaphor to explain pigment buildup over time, and revisit that metaphor during the molting section as a strategic echo. Throughout, you script in reminders to layer in a sound bed of flamingo calls and other ambient sound to enrich tone and provide additional evidence from the field.

Draft a Title

At this stage, draft a working title for your episode. Think of the title as a promise to your audience: it should signal the topic, the tone, and why someone should care enough to press "play." Start simple by combining a hook phrase with a clear description, for example: *"Pretty in Pink: The Science Behind How Flamingos Get Their Color."* You can refine the wording later, but having a working title now helps you stay focused as you research and script. Table 8.3 below demonstrates how our sample episode might look as a two-column script.

Record Audio

Once your script and research materials are in place, your next task is capturing the audio that will become the raw material of your episode. Good audio quality isn't simply a technical concern; it's one of the most important elements of listener engagement. Think of this stage as gathering

TABLE 8.3 Sample Two-Column Episode Script: "Pretty in Pink"

Narration	**Sound**
(Calm, curious tone) *You're standing on the edge of a shallow lake. The air is warm, the water ripples, and in front of you… a sea of pale pink feathers.*	Field recording: water lapping, flamingo calls, distant squawks.
These are flamingos. And believe it or not, they weren't born this color.	Field recording: fade in a single, high-pitched flamingo call.
Their secret? Tiny pigments in their food—think of them as microscopic paint drops—that slowly tint each feather over time.	Sound effect: Soft "shimmer" sound effect under "microscopic paint drops" (metaphor cue).
We'll follow the trail from their favorite snacks—shrimp and algae—all the way to the brilliant pink you see here today.	Field recording: Close-mic sound of shrimp being poured into a metal feeder.
(Interview clip) *"As zookeeper Daniel Ruiz told me when we spoke last month, 'They're basically eating color. The more carotenoids they eat, the pinker they get.'*	Field recording: Ruiz interview
It's not instant. Each molt, the pigments layer in—like adding fresh paint to a canvas.	Sound effect: Light brushstroke sound (optional) under "fresh paint."
By the end of this episode, you'll know exactly why these birds turn pink—and what it says about the ecosystems they call home.	Field recording: Fade back up to flock sounds, then crossfade to gentle music cue.

Note how it builds in sound as evidence and to set the scene, uses metaphors to clarify concepts, uses conversational language, introduces sources, and models the general principles from this chapter.

puzzle pieces: each clip, each voice, and each ambient sound needs to be strong on its own so it can fit seamlessly into the final mix. As a tip, we recommend that after you record your audio, you revisit your script and refine it: replace narrative placeholders such as "insert interview clip here" with the exact dialogue you collected, and confirm sound-effect cues using the actual field audio you captured. This ensures that your script and sound assets evolve together into a coherent and polished episode.

Narration and Interviews

Begin by recording clean, intentional narration. Use a reliable microphone, choose a quiet and acoustically consistent environment (closets work great in a pinch), and record multiple takes of each line so you have options when editing. When recording interviews, always obtain informed consent and explain how the audio will be used in the final episode. Pay attention to mic placement, background noise, and your own presence as an interviewer, as your nods, laughs, and "mm-hmms" can accidentally interfere with the guest's audio. Before wrapping up, record at least 30 seconds of "room tone" (the natural silence of the space) to allow smoother edits later. Once you've captured your raw narration, interviews, and transitions, you have the core building blocks that will shape your episode's narrative arc.

Field Audio

Field audio includes the environmental and atmospheric sounds that bring your story to life: everything from the hum of a laboratory freezer to birdsong from a wetland to the rhythmic whirr of a wind turbine. These sounds function as sonic evidence and help set the scene, reinforcing scientific concepts through texture and mood. When gathering field audio, always capture more than you think you'll need. Collect a range of options: long stretches of ambient sound, short specific moments, and variations in intensity or perspective (e.g., close-up vs. distant). Each clip should be crisp, free of distortion, and recorded with the same care you bring to narration. Later in the edit, these pieces will help you build pacing, transitions, and emotional resonance.

For the flamingo episode, imagine you are working independently with no budget. You record your narration in a makeshift home booth, such as a closet lined with blankets or clothing to dampen echo and reduce outside noise. You use a borrowed USB mic from the campus library or the audio recorder on your phone, making sure to keep it steady and close to your mouth for a clean take. During your zoo trip, you once again use your phone—or a basic field recorder—to capture field audio, recording several long clips of various moments during the trip: flamingo calls, water splashing in the enclosure, the scrape of a shrimp net, and your interview with the zookeeper. Back at home, you upload and review all your clips in Audacity, a free podcast software, choose the cleanest ones, and begin layering them into your working episode file. In your script, you replace placeholders like "insert flamingo sound here" with actual timestamps from your recordings, and you swap in a precise quote from the zookeeper to match the audio you captured. Even on a low budget, these practices help you assemble a rich, polished soundscape that supports the science you are explaining.

Edit and Design Sound

Editing transforms raw audio into a cohesive, engaging story. Choose software that fits your needs: Audacity is a free, open-source editor for beginners handling basic cuts, fades, and multi-track mixes; Hindenburg offers professional features like auto-leveling, voice profiling, and tools for spoken-word storytelling; Descript enables AI-assisted, text-based editing for quick trimming or segment rearranging; and Zencastr records each remote speaker locally for high-quality sound. The key here is to match tools to your project's scope, technical comfort, and time. Beginners may prefer simple interfaces like Audacity, while advanced users or teams might need multi-track editing, AI transcription, or collaborative features such as those in Zencastr. Test two or three options, learn the basics through short tutorials, and commit to mastering one or two—familiarity saves far more time than constant switching.

Editing can be the most time-consuming stage, requiring dozens of small decisions that collectively shape the listener's experience. During editing, tighten pacing, cut filler words ("uh," "like"), and weave in music or effects to reinforce your narrative arc. AI can speed up tasks like noise reduction, auto-leveling, and suggesting cuts, but final editorial judgment is yours—ensuring the story stays accurate, ethically framed, and true to the science. Strong editing sharpens transitions, heightens emotional beats, and builds the rhythm that keeps listeners engaged. Once your episode is cohesive and compelling, you're ready to publish and promote.

For your flamingo episode, again, working independently with no budget means using low-barrier tools. You choose to record and begin basic editing in the free program Audacity, placing the "paint drop" metaphor just before the shrimp-pouring sound so listeners connect concept and sensory detail, and using music lightly to cue transitions. To smooth out final edits, you upload your file into Descript to cut filler words ("uh," "like") and generate an autotranscript for show notes.

Publish and Promote

Formatting and distribution turn your finished episode into something the world can hear. At this point, you shift from creator to ambassador—making sure your work is packaged well, reaches the right audience, and encourages them to come back for more. Here's a shorthand publication checklist to make sure your podcast finds its way to your audience:

Export your audio in the right format. Save your final mix as an MP3 file. This is the standard for podcasting because it balances sound quality with file size. Most editing programs let you set this in the "export" or "save as" menu.

Upload to a hosting platform. Choose where your podcast will live. A podcast hosting platform is simply the place on the internet where your audio files live. The host stores your episodes and generates an RSS feed, a special link that tells podcast apps (like Apple Podcasts or Spotify) when a new episode is available. For a class project, your instructor may recommend a specific option or even keep episodes within the course (for example, uploading an MP3 file to your LMS or a shared drive). If you want your work publicly available, look for a beginner-friendly host with a free or low-cost tier, clear storage limits, and simple instructions for connecting to major apps. You only set up the RSS feed once; after that, every new upload automatically appears in your listeners' feeds.

Create an episode thumbnail. Your thumbnail and title are the first things people see. Make sure they are clear, readable, and engaging.

Include show notes. Show notes should briefly explain the episode's main ideas, highlight guests, and include links or citations to resources mentioned. Better yet, offer a full transcript of the episode for greater accessibility. Remember to credit all music, sound effects, and third-party materials verbally at the end of the episode and in the show notes; ensure accessibility by including transcripts.

Measure and Iterate

Even if you are creating a one-off episode for a class, it is still valuable to reflect on how your choices shaped the listener's experience. If your podcast is published publicly, you can look at simple metrics such as total plays or where listeners stopped listening. If it is shared only within your class, your "metrics" may come from peer feedback, instructor comments, or your own observations during playback. The goal isn't to chase popularity or trends. Instead, iteration helps you understand which storytelling choices worked well and which parts may need more clarity, pacing, or sound design in future projects. For example, imagine that listeners consistently mention how immersive the flamingo field recordings were but seem confused during the pigment explanation. That might suggest adjusting the pacing next time—perhaps breaking the explanation into smaller segments or adding a sound cue to support comprehension. Whether or not you continue producing episodes beyond this assignment, learning to evaluate your work and revise with intention is one of the most valuable skills in podcasting.

TRY IT OUT! HANDS-ON ACTIVITIES

Each of the following activities is designed to help you practice specific aspects of the podcast production process discussed above. Use these activities to build skills gradually, from analyzing professional models to experimenting with sound, scripting, and AI-assisted planning.

ACTIVITY 1: EPISODE DISSECTION AND GENRE COMPARISON

Listen to one narrative science episode (e.g., Radiolab) and one short-form news episode (e.g., Nature Podcast or 60-Second Science). Pause every few minutes and note: How does it open? How are scientific ideas explained? How is sound used (music, silence, tape, ambient noise)? How are transitions and key points signaled? Write a brief comparison: How does genre shape rhetorical choices and audience expectations?

ACTIVITY 2: CRAFT A 60-SECOND TRAILER

Write and record a 1-minute teaser for a hypothetical science episode (in the style of 60-Second Science). Include one metaphor, one deliberate sound cue, and a clear statement of stakes. Swap with a peer. After one listen, ask what they remember most. Revise for clarity, pacing, and memorability.

ACTIVITY 3: EXPLORE YOUR OPENING

For an interview-based science episode, draft three different openings: a narrative anecdote or scene; a paradox, problem, or tension; or a scene using sonic evidence from the field or dialogue from a main character or guest that sets the stakes. Workshop with a partner or in small groups. Which version best establishes stakes and credibility? Revise accordingly.

CONCLUSION

Podcasting's accessible nature, flexible formats, and immersive sound make it a powerful way to bring science into everyday life. When you match genre to purpose, apply strong rhetorical strategies, use AI wisely, and revise based on listener experience, you can create episodes that both inform and inspire. If there is one takeaway, it's this: great science podcasting is equal parts craft and intention. The technical side of microphones, audio formats, and hosting platforms exists to serve a larger goal: engaging audiences in ways that are clear, credible, and compelling. The rhetorical side of podcasting—how you frame your question, design your story arc, and invite listeners into the journey—is what transforms raw information into something memorable and shareable.

In this chapter, we mapped the science podcast landscape, from bite-sized explainers to immersive narrative documentaries, and examined five core rhetorical strategies for audio storytelling: story scaffolding, metaphor, sonic evidence, inclusive framing, and strategic repetition. Popular science podcasts such as *Radiolab*, *Ologies*, *Science Vs*, and *Overheard at National Geographic* illustrate how these techniques make complex science vivid and accessible. We also explored how AI tools are reshaping production by speeding up research, streamlining editing, and facilitating distribution, and we underscored the ethical responsibility to ensure accuracy, transparency, and fair representation when considering using these tools in the podcasting process. Finally, the seven-step podcast process, modeled through the example flamingo episode *Pretty in Pink*, demonstrated how to write for the ear, integrate production techniques with rhetorical strategies, and make informed decisions about publishing and promotion.

Your challenge now is to take these tools and make them your own. Whether you are producing a 60-second teaser, a 10-minute explainer, or a 30-minute narrative episode, the same principles apply: start with a focused question, know your audience, layer rhetorical strategies with purpose, and use sound not as decoration but as a tool for deepening understanding. Done well, your podcast will not only inform; it will stick with listeners, spark curiosity, and maybe even change how people see the world.

REFERENCES

Abumrad, J., Krulwich, R., Miller, L., & Nasser, L. (Hosts). (2002–present). *Radiolab* [Audio podcast]. WNYC Studios. https://radiolab.org

Abumrad, J., & Krulwich, R. (Hosts). (2007, September 21). Sound as touch [Audio podcast episode]. In *Radiolab*. WNYC Studios. https://radiolab.org/podcast/91514-sound-as-touch

Abumrad, J., & Krulwich, R. (Hosts). (2015, June 6). Antibodies part 1: CRISPR [Audio podcast episode]. In *Radiolab*. WNYC Studios. https://radiolab.org/podcast/antibodies-part-1-crispr

Abumrad, J., & Krulwich, R. (Hosts). (2016, August 21). Playing God [Audio podcast episode]. In *Radiolab*. WNYC Studios. https://radiolab.org/podcast/playing-god

BBC. (1948–present). *The Reith Lectures* [Audio podcast]. www.bbc.co.uk

Briggs, A. (Host). (2020, July 21). The canary of the sea [Audio podcast episode]. In *Overheard at national geographic*. National Geographic. https://www.nationalgeographic.com/podcasts/article/the-canary-of-the-sea

Dahlstrom, M. F. (2014). Using narratives and storytelling to communicate science with nonexpert audiences. *Proceedings of the National Academy of Sciences, 111*(Suppl. 4), 13614–13620. https://doi.org/10.1073/pnas.1320645111

Flatow, I. (Host). (1991–present). *Science Friday* [Audio podcast]. Science Friday Initiative; NPR. https://www.sciencefriday.com

Lindenfeld, L., Besley, J. C., Zheng, X., Dudo, A., & Newman, T. P. (2025). *Science communication for scientists: Linking strategy with creativity, practice, and respect* (1st ed.). Routledge. https://doi.org/10.4324/9781003493938

McDonald, B. (Host). (1975–present). *Quirks and Quarks* [Audio podcast]. CBC Radio. www.cbc.ca

McHugh, S. (2022). *The power of podcasting: Telling stories through sound*. Columbia University Press. https://doi.org/10.7312/mchu20876

Ward, A. (Host). (2020, March 10). Virology (Coronaviruses) with Dr. Shannon Bennett [Audio podcast episode]. In *Ologies*. Alie Ward. www.alieward.com

Webster, M. (Host). (2021, August 5). Gonads: Dutee [Audio podcast episode]. In *Radiolab*. WNYC Studios. https://radiolab.org/podcast/dutee

Zukerman, W. (Host). (2019, March 29). Fasting diets: What's the skinny? [Audio podcast episode]. In *Science Vs*. Gimlet Media. gimletmedia.com

Zukerman, W. (Host). (2020, March 5). Coronavirus: Fears and facts [Audio podcast episode]. In *Science Vs*. Gimlet Media. gimletmedia.com

CHAPTER 9

Presentations and Oral Communication

Scientists Take the Stage

WHAT YOU'LL LEARN IN THIS CHAPTER

- A discussion of why you might choose to share science in public out loud
- A description of some of the most important genres of oral communication scientists and science communicators are likely to use: the elevator pitch, media interview, and conference talk
- Core principles for engaging listening audiences through structure, delivery, and visual support
- A process for distilling complex written arguments into an engaging, concise oral performance

INTRODUCTION

On a stage in Vancouver in April 2023, professor of computer science Yejin Choi stood before a large TED audience to explain one of the great paradoxes of artificial intelligence. Large language models like GPT-4, she explained, can pass the bar exam and ace medical licensing tests—achievements that require synthesizing vast amounts of specialized knowledge. And yet, these same models struggle with questions a 7-year-old could answer.

DOI: 10.1201/9781003520887-11

When asked, "If I have a 6 liter jug and a 12 liter jug, how do I measure 6 liters?" the model spit out "elaborate nonsense," Choi reported, eliciting a collective chuckle from her listeners (Choi, 2023). The model was "incredibly smart and shockingly stupid"; its reasoning impenetrable, its environmental burden dangerous, and its engineering costs so high they concentrate power in the hands of a few tech companies (Choi, 2023). In just 13 minutes, Choi made the promise and peril of cutting-edge AI research accessible to a public audience using analogies, carefully chosen examples, and strategic repetition.

While originally founded as a technology conference in the 1980s, TED now hosts talks on a variety of themes ranging from climate change to innovation and personal growth. Recordings of the talks are posted online for free, reaching billions of people around the world every year (TED Conference, n.d.). Over 2 million people have watched Choi's TED Talk, likely a significantly larger number than those who read her research papers in computational linguistics journals.

Since the rise of the scientific conference in the nineteenth century, oral communication has played an increasingly important role within modern science by strengthening scientific communities and disciplines. At conferences, scientists share their research through formal talks, poster presentations, and informal conversations with colleagues, standardizing knowledge and fostering new collaborations (Alder, 2012). Spoken words have shaped the public understanding of science, too—both in conferences organized deliberately to reach a general audience, like TED, and in many other spoken genres. For example, Rachel Carson, marine biologist and author of *Silent Spring* (1962), took the legal stage, testifying before the United States Congress about the harms of pesticides. During the COVID-19 pandemic, Anthony Fauci, then director of the National Institutes of Allergy and Infectious Diseases, regularly briefed the public on the emergent virus. In these cases, as science was being communicated in public, it was also *being shaped,* as questions and comments invited new articulations of research motives and findings.

Consequential science communication such as that by Choi, Carson, and Fauci requires preparation and practice, which has long been theorized. As early as the first century BC, the *Rhetorica ad Herennium* synthesized classical rhetoric, describing the steps communicators move through to reach an audience in what are known as the canons of rhetoric: from invention, arrangement, and style to memory and then delivery (Horner, 1988).

In the ancient world, oral communication was ephemeral, heard in an instant. Today, however, both scientists and science communicators must attend to a distinct media environment, what we refer to in Chapter 1 as a media ecology. As we've seen, if recorded, talks may persist indefinitely online and may reach global audiences. A single well-crafted presentation can also be cited in news articles and excerpted across social media. When you communicate out loud today, you may be crafting something that's highly durable or something that cannot be paused or rewound. By exploring the distinct challenges and opportunities of communicating science out loud, this chapter will help you explain the science you care about in 3 or 30 minutes to a listening audience.

OPPORTUNITY AND CONSTRAINT IN ORAL GENRES

Scientists and science communicators engage in oral communication for reasons both practical and intellectual: to educate, spark interest, learn, seek funding, or advocate. Science writer Annaka Harris, for example, published the book *Conscious: A Brief Guide to the Fundamental Mystery of the Mind* (2019) with Harper, a trade publisher, and then developed a follow-up audio documentary, *Lights On: How Understanding Consciousness Helps Us Understand the Universe* (2025). The audio documentary makes audible Harris's in-depth conversations with astrobiologists, physicists, psychologists, and philosophers. This oral genre extends the reach of her written work and gives her audience direct access to the voices of the experts she's interviewing about controversial theories.

Today, science communicators like Harris can select from an ever-evolving set of genres with an oral component. In 1996, the New London Group, a community of international academics, observed the rise of what they called *multimodality*, or meaning-making from linguistic, visual, aural, gestural, and spatial designs (Cazden et al., 1996). When you give a talk, for example, you are communicating multimodally because you are using multiple modes—spoken words, visuals, embodied rhetoric, and a position within the space of the room—all of which, in combination, contribute to the meaning you make with your audience. The concept of multimodality helps us see the importance of modes beyond the written word to human communication. Coupled with the proliferation of internet-enabled platforms, multimodality has provided science communicators with more opportunities—and more genres than ever—to participate in culture since the 1990s. To take a prominent example of another genre that has an oral component: when the online video-sharing platform YouTube made it possible in 2005 for anyone with a laptop, a camera, and video-making software to upload videos, many science communicators jumped at the chance to create explainer videos. These videos explain scientific concepts to broad audiences by combining voice-over narration with a variety of visuals. They are a good example of a genre where oral communication is imbricated with other modes of communication.

Our coverage of oral genres in this chapter is not exhaustive. We have chosen to focus on the elevator pitch, media interview, and conference presentation—oral genres with an in-person component—because in our experience, they are the most necessary to science communicators. Podcasts are another oral genre that has become very popular in the last few years, so popular that we elected to dedicate Chapter 8 to them. They may be highly scripted audio stories or free-form conversations. Regardless, listeners tune into them much as they would a radio channel, but typically as their schedules permit thanks to streaming platforms. Podcasts make audio science highly accessible. They also call attention to some of the limitations of oral genres. In oral communication, even if your work is based upon extensive research, it can sometimes be more challenging to convey the provenance of your ideas and sources (especially in formats that do not allow for slides).

Yet oral genres make scientific work available in ways that other forms of communication cannot. Publications often exist behind paywalls, and their language and structure may further complicate access to the findings. Effective oral delivery, by contrast, takes the needs of the listening audience into consideration, working to make the science both understandable and relevant. Moreover, a responsive speaker demonstrates human judgment, creativity, and contextual understanding—the very qualities Choi highlighted in her TED Talk when she explained that AI lacks "common sense," the ability to reason about everyday physical and social situations that humans acquire through lived experience.

Throughout this book, we promote an understanding of science communication as social engagement with dynamic audiences. Interviews and live talks are fantastic opportunities to understand both what matters to your audience and what confuses them. When you take the stage, you experience the audience's interest, boredom, or skepticism almost immediately. Furthermore, their questions and critiques often shape your own thinking, research, and future communication choices. Nowhere is science communication more of a two-way street than in informal and formal oral communication. This is a wonderful opportunity—and also a risk—for science communicators.

The Elevator Pitch: Research in 90 Seconds

The elevator pitch—a concise, persuasive articulation of your research that takes no more than 90 seconds, about the time it might take to ride an elevator—serves many professional needs. You might deliver one at a conference networking event, in a chance encounter with a potential research collaborator, in front of your own research poster, or when an editor asks about your latest findings or story. The goal is to stimulate curiosity and create opportunities for deeper conversation.

The elevator pitch is prominent in business because entrepreneurs need to capture investors' attention in brief encounters. In scientific settings, or in academic settings more broadly, the elevator pitch is most likely to take place in the context of a conference or other gathering where random, in-person encounters are likely. You might also strategize to deliver an elevator pitch: after a panel at a conference, for instance, you might plan to approach a speaker to introduce yourself and offer a brief spiel about a project, research, or publication idea. In the general public, it serves a similar function: you have perhaps 90 seconds to make someone interested in the science that matters to you. The pitch isn't comprehensive; rather, it opens a conversation by highlighting what's most compelling about the research.

Effective elevator pitches follow a predictable structure. They begin with a hook that connects the topic to something your listener already cares about or finds intriguing, and a brief introduction to who you are, if necessary. Then, they briefly describe the research question and perhaps the approach. If there are findings, a pitch will describe them and argue for their significance. They often end by inviting dialogue, or future correspondence.

Developing an elevator pitch serves another crucial function for scientists and science communicators alike: it helps you clarify your thinking. The process of distilling a complex research project into 90 seconds of accessible language forces you to identify what matters. As composition research has shown, speaking ideas aloud helps many writers move beyond dense academic language and focus on the concepts that animate their work (Elbow, 2011). If you can't explain your research concisely to a nonspecialist, you may not yet understand it well yourself.

A TEMPLATE FOR AN ELEVATOR PITCH

To prepare for an elevator pitch, first watch at least three given by other people. The Massachusetts Institute of Technology, for example, publishes its pitch competitions on YouTube. Take notes on what makes a pitch successful and the pitfalls to avoid. Be specific. What do you admire? What strategies do you want to try out, practice, and make your own? Conversely, are there particular rhetorical moves that do not seem suited to your goals or audience? You can choose not to emulate those and find alternatives.

The following template can then help you craft your own elevator pitch. Write out 1–2 concise sentences for each prompt:

The hook establishes the stakes immediately—why should anyone care about this research?

Next, briefly describe your *research question and approach*. What problem are you addressing? What gap in knowledge are you trying to fill, and how?

The *significance statement* answers the "so what?" question. Why does this research matter beyond your narrow subfield? If you get stuck on this question ask yourself what *The Craft of Research* recommends: What would be the consequences for the public if no one did this research or shared it? What would they lose out on understanding? (Booth et al., 2024).

Consider offering *a concrete example* to illustrate your work. Examples make abstract concepts tangible.

You can end in one of two ways: either by articulating the *contribution* (what's new?) or by *inviting dialogue*. Asking a question that engages your listener transforms your pitch from a monologue into a conversation.

Practice saying this 8–10-sentence script out loud a number of times to find ways to strengthen your language and build smooth transitions. Time yourself so that you stick to 60–90 seconds!

The Media Interview: Science on Camera and Microphone

Media interviews serve different purposes. When journalists contact scientists for comment on breaking news or emerging research, they're seeking expert perspectives that can help public audiences make sense of complex developments. A climate scientist might be asked to explain new temperature data for a television news segment. A materials engineer might discuss the failure of a bridge structure for a radio program. These interviews translate specialized knowledge for broad audiences, often under significant time pressure, and they often emerge from other publications or exigent science communication occasions. For instance, if you publish an op-ed in a regional newspaper, you might be asked by a local TV news station to give an interview following up on your perspective.

When a journalist contacts you for an interview, they're working under different constraints than you are. They need accessible explanations quickly, they're producing content for audiences with limited technical backgrounds, and they're often working within rigid time or space limits—a 90-second television segment, a 3-minute radio piece, or a 300-word online article.

Experts in their fields may be called to weigh in on current events or emergent problems. Some of those rhetorical situations are addressed in Chapter 6: *Scientific Opinion Writing*. However, media interviews often emerge in response to other science communication occasions described in this textbook. In particular, news sources frequently respond to one another and across mediums; this transpires regionally as well as nationally. For instance, if you publish a timely op-ed in a regional newspaper, a local television station might contact you asking you to give your position in an interview as well.

Going on television or the radio can be both an exciting and an anxiety-inducing experience. If asked to give an interview, it's perfectly acceptable to ask if you can have the questions in advance if there's time, although not all venues will be willing to provide this information. If you are unsure whether you want to participate in an interview on a short timeline, you can always decline. Requested media interviews are opportunities, and you should only take advantage of them if you feel confident that the venue and interviewer will help you positively communicate your message.

During the interview, you should keep your responses direct and concise. If a question surprises you, it's fine to ask for clarification. If you feel like a question is outside your area of expertise, or that you are being asked to speculate on something potentially controversial, you can always decline to answer with a simple, friendly, "I think I'd prefer not to comment at this time."

One important caveat regarding media interviews: it's important to recognize that appearing in a given forum provides a tacit nod to the legitimacy of that forum. In an age where our news ecosystem is frequently hyperpartisan, appearing as an expert on a network or show with a strong political bias lends your credibility to that forum. Even if your comments challenge the forum's prevailing politics, your presence can be construed by outside observers as a form of participation.

AN ACTIVITY TO PREPARE FOR A MEDIA INTERVIEW

Use this activity together with another science communicator to help you prepare for a media interview. Find a partner who also anticipates media interest and who can ask you questions. You can make some suggestions about approaches to the questions, but ideally, they will choose the wording so that there's an element of surprise for you. Give answers that would be appropriate for a public audience, specifically that of a morning television news program. In return, you should also ask your partner questions.
This activity will help you distill the core message of your science, practice being both an interviewee and an interviewer, and reflect on the elements of a good interview.

1. Make sure you each have a way to record the interview—a phone (using a voice memo app or similar) is easiest, though you could record onto your computer as well. You will record both the conversation when you are the interviewer and the conversation when you are the interviewee. You'll listen back to the interviews to help you reflect.
2. Find a quiet place to record your interview with your partner.
3. Conduct the interview. Decide who will take which role (interviewer or interviewee) first. Then, the interviewer should interview the interviewee for approximately 3–5 minutes. Remember to *listen* to what your partner is saying so that you can ask specific follow-up questions. You want to have a conversation about the science and what's exciting or interesting about it.
4. After the first interview, switch roles and repeat. Keep an eye on the time so that both partners have equal turns in each role.
5. After you've completed the interviews, listen to both recordings and either talk about or write in response to these prompts: What were the challenges in engaging in the interview *conversation*? As the interviewer, how well did you handle asking questions? As the interviewee, were you able to get across the most important ideas in the science?

The Conference Talk: Translating Written Arguments for Listening Audiences

Conference presentations represent perhaps the most common form of oral scientific communication, as they allow scientists to share findings with diverse audiences while refining their thinking through the act of explanation. Why do conference presentations count as public presentations? Because presentations that are made at colleges or universities are often open to the public. And even at highly technical conferences, you may be addressing people very new to the field, people in other subfields, or industry or policy partners who are eager to learn from a communicator who can make new science understandable and exciting. Other conferences are less technical. The Stanford Medicine "Big Ideas" series, for instance, invites researchers to present their work to audiences that include clinicians, scientists from other disciplines, administrators, and interested community members. "Conferences" like TED are deliberately organized with and for a broad public. Therefore, we use the term *conference presentation* somewhat loosely to refer to a prepared talk of any length that often includes visual slides.

If you are preparing a conference talk, you will want to reflect carefully on the kind of audience you are presenting to—whether specialist, generalist, or somewhere in between. As we've seen, sometimes presentations are primarily for scientists from the same discipline (like at a disciplinary conference), sometimes they are meant to engage researchers from different parts of the academic community (students and people from other disciplines), and sometimes they are directed primarily at the public. No matter who you are talking to, your audience will want to know the research question, findings, and, perhaps most importantly, the implications. Beyond that, you will need to tailor your talk further depending on who you are talking to. In general, disciplinary specialists want to hear about careful methods and a contribution to the field. Broader academic audiences want to understand the conceptual implications of the study beyond a narrow subfield. Public audiences often want to know how the science affects their lives or society.

Preparing a conference talk forces scientists and science communicators to identify what matters most in their work and why nonspecialists should care. Interdisciplinary inquiry and problem-solving are facilitated. But unlike written papers, which readers can revisit and reread, oral presentations move at a pace you control but listeners cannot. And unlike brief elevator pitches or media interviews, conference talks allow the development of complex arguments—typically speakers are permitted 15–25 minutes. But this extended time brings its own difficulties. How do you maintain audience attention and comprehension throughout? The strongest conference presentations don't simply report findings; they tell a story of scientific inquiry. They have a clear narrative arc, moving listeners through questions, methods, findings, and implications in a way that feels coherent.

Our section on structure below says more about how to achieve this coherence.

While the subgenres of the elevator pitch, media interview, and conference talk are distinctive, they all have in common their ability to reach a listening audience. But what does a listening audience need? Let's look next at that.

PRINCIPLES OF EFFECTIVE ORAL COMMUNICATION

Know Your AGE: Audience, Goal, Environment

No matter its length or format, oral delivery requires attention to three fundamental elements: Audience, Goal, and Environment (Oral Communication Program, n.d.).

Audience: Successful oral arguments take into account what the audience already knows and what it still wants to learn, its hopes and its fears, as well as its values. Speakers at specialized conferences assume a different level of background knowledge than those at a public lecture. When Yejin Choi presented at TED, she tailored her content for nonspecialists in her field—she didn't assume her audience knew what "symbolic knowledge distillation" meant, but she also didn't condescend or oversimplify to the point of inaccuracy (Choi, 2023). She also knew that audiences come to TED conferences to hear good news: how problems can be solved. So, her tone stays confident and upbeat throughout, even as she acknowledges that computer engineers are facing formidable challenges as they strive to build useful artificial intelligence.

Finally, she takes rhetorical risks to make her technical subject matter accessible to a general audience. Most prominently, she uses personification, a kind of metaphor, to help her audience relate to AI ("it can ace the bar exam"). But that very move is also what has some people overrelying on AI for reasoning and connection, because they take the metaphor of human intelligence literally. Addressing a talk to a particular audience involves risks that you will want to weigh carefully. In Choi's case, she balances the optimism of personification with critical distance: she figures AI as "a new intellectual species" (Choi, 2023).

Goal: Your goal for your speaking shapes everything from content selection to level of detail. Choi not only wanted to explain the paradox of her talk's title—how large language models can seem both stupid and stunningly intelligent—she also wanted to make the case for the painstaking work of teaching models "common sense" (she concedes people in her field think it is a moonshot) and, implicitly, for funding to research universities that might pursue this kind of work. Because these goals are varied, her talk cannot address a lot of specifics.

Environment: All talks take place in a material space. A small conference room may encourage questions and even intimacy whereas a large auditorium requires projection, and often, a commanding presence. Choi is in the latter, so she must use all the tools at her disposal to reach everyone in the auditorium: a mic, a position at the front of the stage, effective visuals, intentional body language, and a confident delivery.

Your Script: Making Complex Ideas Accessible to Listening Audiences

It may seem counterintuitive, but you can write to be heard. Speakers often craft a script to help them collect their thoughts and find the best words to reach a listening audience. Adapting technical material to engage listening audiences requires a few different strategies from those needed to engage generalist readers, though much of what we say in Chapter 2 on rhetorical tools can also be applied to writing that's meant to be spoken.

Listeners process information in real time and cannot pause to look up unfamiliar terms. This means you must integrate definitions and explanations seamlessly. When you introduce technical vocabulary, define it immediately and concretely. Rather than immediately saying, "Optimized learning occurs in the zone of proximal development," explain: "Humans learn best when they are neither bored by what they already know nor so challenged by new material that learning feels impossible." The technical term can come after with appropriate credit: "This window for learning is what Lev Vygotsky has called the zone of proximal development." When you must use an acronym, speak out the letters so your audience can hear each one and then say what they stand for in a contextualizing sentence or two. For example: "SSRIs, Selective Serotonin Reuptake Inhibitors, increase the level of the feel-good hormone serotonin in the brain, providing psychiatrists a generally safe treatment for depression."

For an audience interested in science, we recommend an *elevated conversational style*. This doesn't mean being casual or imprecise—it means writing sentences in your script that are designed to be heard rather than read and that also convey respect for the specialist content and language. Compare these versions by reading them out loud and really listening to their impact:

Written: "The amelioration of coastal erosion requires multifaceted interventions addressing both anthropogenic and natural etiological factors."

Oral: "To slow coastal erosion, we need solutions that address human activities and natural causes."

The second version uses shorter sentences, simpler vocabulary, and the active voice. It sounds like intelligent conversation rather than like reading a journal article aloud, which would quickly bore your listening audience.

Repetition serves crucial functions in oral communication that would seem redundant in writing. Key concepts benefit from being expressed multiple ways. You might explain an idea, then illustrate it with an example, then return to it later in different words. Strategic repetition helps listeners who may have momentarily lost focus or who need multiple exposures to grasp complex material. Early in her talk, Choi analogizes large language models to Goliath, the famous giant of the Bible who was defeated by the smaller and wilier David. These models are unwieldy and unwise, her analogy suggests, and later in the talk she repeats the comparison when she wants to disparage the current training approach (the building of ever-larger models) to make space for her own method, which is smaller and more targeted—more like a David in its approach to problem-solving.

Structure: The Foundation for Listener Comprehension

Your audience's ability to follow your oral argument depends heavily on its structure (above we used the classical term *arrangement*, but today it's more often referred to as *structure* or *organization*). Unlike readers who can look back to remind themselves where you're going, listeners must hold your entire argument in working memory. This makes explicit and logical structure essential.

All talks require a goal, as we've seen. But you need to get your audience to care about your motive, too, which you can do with a specific structure. Try starting with what rhetoricians call the "stasis" or "status quo"—what we already know about your topic (Booth et al., 2024). In Choi's TED Talk, she established this by explaining that large language models have become remarkably capable at certain kinds of writing tasks requiring specialized knowledge. This gave her audience a foundation for understanding what came next.

Then, introduce the "disruption"—the problem, gap, or question that motivates your research (Booth et al., 2024). Choi's disruption was the paradox: these same powerful models fail at common-sense reasoning. The disruption creates tension that your talk will resolve. Make the consequences of this gap clear. Why should your audience care about this problem? Choi explained that without common sense, AI systems make dangerous mistakes, like providing unsafe advice.

Your research then becomes the "resolution"—how you addressed this problem (Booth et al., 2024). Note that offering a resolution doesn't mean you must claim to have solved the problem. Often in science, the resolution poses new useful questions given what was learned, charts a new method, or verifies the findings of other research to solidify scientific consensus. The significance and resolution can take many forms and will vary according to what you want to argue and your audience's expectations. What is important is that your talk ends by directly addressing the tension, problem, or gap that was raised at the outset.

For example, Choi described her work developing large language models that can reason about physical and social situations to help AI systems perform something closer to common sense. This narrative structure—stasis, disruption, resolution—mirrors the psychology of how humans process information. It's fundamentally a story structure: "Once upon a time, we understood things this way... but then we discovered this problem... and here's how we're responding to it."

Choi then uses another classic rhetorical strategy to defend her resolution to the audience: tricolon, or the rule of three. This can be an especially effective approach for listeners who cannot hold too many ideas in their heads at once. Choi announces that to achieve AI with common sense, scientists must take inspiration from three lessons from Sun Tzu's *The Art of War*: know your enemy, choose your battles, and innovate your weapons (Choi, 2023). Naming these lessons toward the top of her talk helpfully *forecasts* its structure; then, as she moves through each subsection, she repeats each of the lessons verbatim so that the audience is aware of where they are in the talk. Choi reinforces the repetition in her oral argument with slides that also articulate the lessons in the fewest words.

IF YOU'RE LOOKING TO LEVEL UP YOUR SPEAKING SKILLS, STUDY EXCELLENT SCIENTIFIC CONFERENCE TALKS

1. Watch Choi's TED Talk for yourself. Then, compare it to one of her scholarly papers. How has she made the science relevant and exciting to the public while representing it accurately? If you were on that TED stage, what would you do differently, if anything, and why?
2. Reflect on some of your favorite public speakers. What do they do that makes you want to listen to them? What helps you understand complex ideas? Make a list of these strategies and then try folding them into your next oral presentation.

Visual Support: Slides as Argument

If technology permits, you may choose to create a slide deck to complement your oral argument. As you design your slides, think carefully about how your oral argument and visual argument interact. Well-designed slides enhance comprehension and retention, but poor slides distract the audience, interrupt their listening, and even potentially confuse them. The fundamental principle is this: slides should support your oral argument, not replace it. The slides are for your audience: they are there to support your audience's understanding, not to help you remember what to say. While there are many wonderful books on slide design—we've listed some of our favorites below in the further resources section—here we briefly provide our top tips:

Especially in longer talks (10 minutes or more), provide a roadmap and consider returning to it. Early in your presentation, show your audience how you've arranged your talk—its structure as articulated in your subtitles or main questions. Then, as you transition between sections, return to this roadmap slide, or a version of it, to highlight where you are now. This technique, sometimes called a "location slide," prevents listeners from getting lost. A simple roadmap might show the following:

- *Background*: What we knew about ocean acidification
- *Problem*: Recent discoveries that complicate our understanding
- *Our Approach*: New methods for measuring pH changes

- *Findings*: What we learned
- *Implications*: What this means for marine ecosystems

Each time you move to a new section, briefly show the roadmap again with the current section emphasized through bolding or highlighting. This creates visual coherence and helps audiences track your progress through complex material.

Have a clear point for each slide. Every slide should serve a specific rhetorical purpose in your argument. If you cannot articulate why a slide matters, delete it. Each slide should answer one of these questions: What do you want your audience to see—and why? What concept or point are you illustrating? What evidence are you presenting? *The design of each slide will then emerge from its specific purpose—and location—in your talk.* For example, you will want to use a different slide layout for your title than you do for the presentation of a graph that serves as important evidence. Some slides can even go blank. Offer a black or white background screen during portions of your talk where you want full attention on your spoken words and gestures.

Less is always more. When slides contain many complete sentences or a dense paragraph, audiences face an impossible choice: read the slide or listen to you. They cannot do both effectively. People read faster than you speak, so they'll finish reading before you finish talking, then wait (impatiently) for you to catch up. Especially when you're presenting to a generalist audience, limit slides to key words or phrases, a couple of tight sentences, or compelling images (see Figures 9.1–9.3).

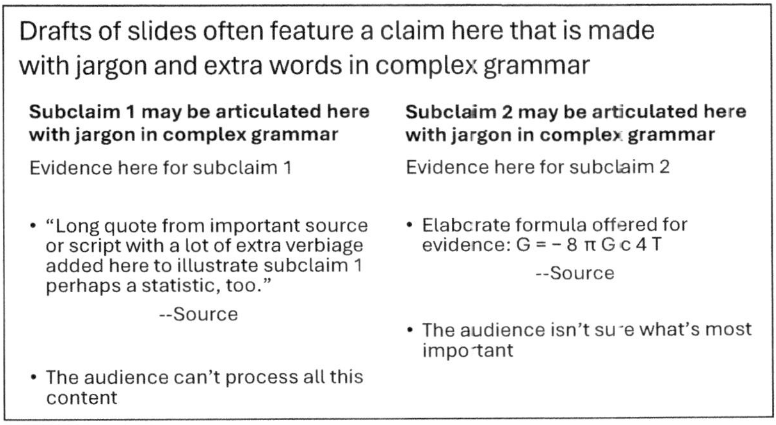

FIGURE 9.1 First drafts of slides are often text heavy and difficult to interpret.

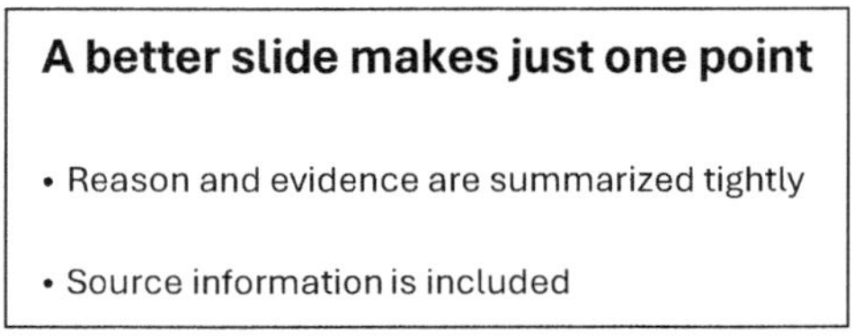

FIGURE 9.2 A better slide is pared back to its essentials and makes just one point.

FIGURE 9.3 A better slide makes just one point and uses a visual to help the audience grasp the argument and evidence quickly.

Reveal information as it is needed. If you have three bullet points you want to make on a single slide, don't show all three simultaneously. Bring up each point as you discuss it. This keeps your audience's visual attention synchronized with your verbal content. You might dim or gray out points you're not currently discussing, bringing each into full color or brightness as you address it.

Use the power of images strategically. A compelling photograph, diagram, or illustration can convey information that would take paragraphs to explain in words. But images must be high quality, clearly visible from the back of the room, and directly relevant to your argument. Avoid decorative images that serve no rhetorical purpose. When showing scientific photographs, diagrams, or data visualizations, ensure all text is large enough to read. Better yet, recreate complex figures from papers as simpler versions optimized for projection. A graph that works well in a journal article often contains too much detail for a slide. Be sure to cite the sources of your images on the slide itself.

Apply design principles for visual coherence. Four principles improve nearly any slide and slide deck:

- *Contrast* ensures important elements stand out. Use size, color, or weight to emphasize key information.
- *Repetition* creates unity. Use consistent fonts, colors, and layouts throughout your presentation.
- *Alignment* creates clean, professional-looking slides. Align text and images along invisible grid lines rather than scattering them randomly.
- *Proximity* groups related items. Place related information close together and unrelated information farther apart.

Choose colors and fonts carefully. Ensure strong contrast between text and background—not just for visibility but for accessibility. Color vision deficiency is not uncommon. Avoid relying solely on color to convey information; use shapes, patterns, or labels as well. Colors also carry emotional weight: cool colors (blues, greens) feel calm and professional; warm colors (reds, oranges) draw attention and can signal urgency or importance.

For typography, sans serif fonts (like Calibri or Arial) project more clearly than serif fonts (like Times New Roman). Use no more than two complementary fonts in any presentation—perhaps one for headings and another for body text. Font size matters enormously: text smaller than 24 points often becomes difficult to read from a distance.

Match your data visualization to your message. If you want to show trends, relationships, or comparisons, visual displays work better than tables. Bar charts excel at showing comparisons; line charts reveal trends over time; scatter plots show relationships between variables. Pie charts should be reserved for simple compositions—and never used for comparisons, as people interpret angles variously. Whatever format you choose, show only what your audience needs. Remove grid lines, labels, or data series that don't serve your specific point. For more on visualizing data, see Chapter 10.

Delivery: Voice, Gesture, Presence

Even brilliant content delivered weakly will lose audiences. Effective delivery doesn't mean theatrical performance; it means purposeful, authentic presentation. Passion for your subject, communicated through your delivery, makes the science compelling. As Winston Churchill observed, "Before [the orator] can inspire [the audience] with any emotion he must be swayed by it himself. Before he can move their tears his own must flow. To convince them he must himself believe" (quoted in Koegel, 2007).

Vocal variety: Vary the pace, pitch, and volume of your voice to maintain interest and emphasize key points. A monotone voice makes even exciting research sound boring. Slow down for complex or crucial information, allowing your audience time to process. Speed up slightly when moving through familiar background material. Pause strategically—silence gives listeners moments to absorb what you've said and builds anticipation for what comes next.

Pitch variation conveys emotion and emphasis. Raising your pitch can communicate enthusiasm or signal a question; lowering it suggests seriousness or finality. Volume changes create drama—speaking more softly can draw audiences in and force them to lean forward and listen carefully, while speaking more loudly emphasizes critical points.

If the room is large, be sure everyone can hear you; if not, request a microphone to make your delivery accessible to all. Moreover, if technology is available to enable live captioning, make sure to activate it from your slide deck before beginning your presentation. Research has shown that captioning a video improves comprehension, attention, and memory, particularly for persons watching videos in their non-native language and for persons who are Deaf or hard of hearing (Gernsbacher, 2015; Vanderplank, 1988). We can generalize these findings to presentations because the underlying cognitive mechanism is the same: the listening load is reduced when a visual channel is made available.

> Practice your vocal variety by reading a favorite passage aloud multiple times, exaggerating each element—pace, pauses, volume, and pitch—to expand your natural range. Record yourself and listen critically to identify when you're most engaging.

Gesture and movement to connect with your audience: Look at your audience as much as you can, using your notes discreetly. When you look down at your notes or back at your slides, you potentially break your connection with your audience. By contrast, eye contact allows you to gauge your audience's engagement and understanding, facilitating an informed response.

Your hands function as "traveling visual aids," helping you tell your story visually (Koegel, 2007). Natural gestures that align with your words enhance communication. Use your hands to indicate size, show relationships, or emphasize points. Avoid repetitive, unconscious gestures (such as swaying, fidgeting, hand movements like "rowing," or the defensive, vulnerable "fig leaf" position) that distract from your content.

Your posture can also help communicate your engagement with the material. Even if you are sitting, your body language can indicate your desire to connect with the audience; for example, by leaning forward toward the audience. If you are standing, keep your body tall and your head still—but don't remain rigidly fixed in place. Purposeful movement, such as stepping closer to emphasize a point or moving to different areas of the stage to signal transitions, enhances engagement and can often help you feel more at ease.

Managing status through delivery: Theater director Keith Johnstone's concept of *status* helps explain how delivery choices affect audience perception (Johnstone, 1981/1992). High-status delivery—a direct gaze, complete sentences, purposeful movement, and minimal filler words—projects confidence and authority. Low-status delivery—an averted gaze, fidgeting, incomplete sentences, and frequent "ums" and "ahs"—can make you seem uncertain or unprepared. While you generally want to project confidence, there may be strategic moments to lower your status slightly—for instance, when acknowledging the limitations of your research or when inviting audience collaboration and questions. Being aware of these signals helps you make intentional choices about how you present yourself.

Communicating passion authentically: Audiences respond to speakers who genuinely care about their subjects. But passion doesn't mean histrionics or forced enthusiasm—it means allowing your authentic interest and investment in your research to come through in your delivery. Think about why this work matters to you. What questions keep you up at night? What discoveries excited you? Let that energy inform your delivery, including purposeful embodied rhetoric. As communication consultant Timothy Koegel notes, your body language sets the tone of your presentation before you even begin speaking (Koegel, 2007). If you seem reluctant to speak or uncommitted to your own material, your audience will begin to disengage too.

Managing anxiety: Most speakers experience nervousness before and while they present. Matt Abrahams, a lecturer in organizational behavior and communication consultant, explains that public speaking anxiety stems from our perception of being evaluated in real time, which triggers physiological responses: increased heart rate, rapid breathing, muscle tension, and racing thoughts (Abrahams, 2016). Rather than trying to eliminate anxiety entirely—which cannot be done for most of us—develop strategies that help you manage it.

Cognitive reframing helps reduce anxiety by changing how you interpret the speaking situation. Instead of viewing your presentation as a high-stakes performance where you might fail, reframe it as a conversation where you're sharing information with interested people (Abrahams, 2016). This shift in perspective can reduce the perceived threat.

Practice matters enormously, but how you practice matters too. We suggest you practice under conditions similar to your presentation circumstances if you can. Practice standing up if you'll present standing. Rehearse with the technology you'll use. If possible, practice in the actual room where you'll present. This situational practice reduces novelty and uncertainty, two major anxiety triggers (Abrahams, 2016). Visualization is a kind of mental rehearsal that can also reduce anxiety. Spend time imagining yourself delivering your presentation successfully—see yourself speaking clearly, notice your audience engaged and nodding, and feel yourself confident and composed. This mental practice builds familiarity with the situation and confidence (Abrahams, 2016).

Before and during speaking, you can address the body's stress response directly. Deep breathing—particularly extended exhalation—activates the parasympathetic nervous system, which calms your body. Before presenting, try breathing in for four counts, holding for four counts, and exhaling for six to eight counts. Repeat this several times (Abrahams, 2016). While presenting, don't forget to breathe! You might feel like you can't afford a breath, but when you take a beat, you'll deliver your next sentence or paragraph more clearly, and give the audience time to collect their thoughts, too.

Finally, try to remember that mild anxiety can enhance performance. The energy anxiety creates can help you bring dynamism to your presentation. You don't want to feel so relaxed that you seem apathetic or sluggish to your audience!

Memory and Extemporaneous Delivery

You face a choice about how to remember and deliver your content. Reading verbatim from a script makes you sound like you're reading—you lose eye contact and often vocal variety and the ability to respond to audience reactions as well. On the other hand, pure improvisation risks incoherence, rambling, or forgetting crucial points. The solution is extemporaneous delivery: delivery that is carefully prepared but not scripted, allowing you to speak naturally while ensuring you cover essential material.

We strongly recommend you write out your talk in full sentences, then distill it into an outline or bullet points. Practice delivering from these notes until you can speak fluently without reading. Your slides themselves can also serve as memory aids if used discreetly; as you display them, each slide will trigger the language and ideas you want to share.

Another approach is to memorize your opening, your key transitions, and your conclusion. These structural elements benefit from precise language. For the body of your talk, know your content thoroughly enough that you can explain it naturally, perhaps varying the exact words each time you practice. You might even find better ways of expressing your most important ideas!

THE ETHICAL DIMENSION OF ORAL SCIENCE COMMUNICATION

Speaking about science publicly carries responsibilities. You represent not just yourself but also your field and the scientific enterprise.

When your oral communication is inclusive and accessible it shows your audience that you care about the impact of your speech—that their needs and concerns matter to you. Using language and analogies that respect your audience, rather than marginalizing them, is paramount. We suggest you return to Chapter 3 for ways to think further about how to make your oral presentations inclusive.

When you're speaking, it may be tempting to rely on appeals to emotion. You can insert a quick joke more easily into a speech than into most written genres, and that move can be a great way to recapture your audience's attention or to make a point. But don't be tempted to exaggerate or oversimplify findings to make your presentation more dramatic. If the evidence does not support your claims, then people are unlikely to listen to the rest of what you have to say. In fact, it's wise to communicate the limitations and confidence levels of your findings. In Choi's TED Talk, she was careful to explain that many in computer science are skeptical of her research goals and that developing common sense in AI remains a distant possibility. Acknowledge what you don't know; your humility will win trust.

As you speak, credit collaborators, prior research, and sources of funding. In written work, this happens through citations; in oral presentations, you create a verbal tag. These attributions can build the credibility of your claims by referencing out loud the journal title, the recency of the publication, the institutional home of the researchers, or their methods. For example, you might say, "In their longitudinal study of diabetes that came out last month in *The Lancet,* University of California researchers found …."

Though there is no consistent style guidance on citations on slides themselves, we suggest that for a conference presentation, you both make your sources explicit on the relevant, individual slides and include a final works cited slide, especially if your audience is deeply interested in research. If you have not generated them yourself, the images and graphics in your slides should also be cited.

In most contexts, it is appropriate to disclose relevant relationships that might influence your work or presentation: funding sources, financial interests, or professional connections that audiences should know about. Disclosures usually come at the start of a presentation both out loud and in writing on the slides. They can be brief but should be unambiguous.

CONCLUSION

The skills and strategies this chapter has described—narrative structure, accessible language, strategic visual support, and purposeful delivery—require practice to develop. Like any form of communication, oral presentation improves with feedback, revision, and experience. But the investment pays dividends. Whether you're explaining your lab work in 90 seconds to a potential collaborator or presenting your latest findings to a conference audience, you're doing more than transmitting information. You're establishing credibility and building relationships.

As artificial intelligence becomes increasingly capable of generating written text, the value of oral communication may rise. Because AI introduces bias and inaccuracies and casts doubt on the existence of a human author, an AI-written stigma has emerged (Lermann Henestrosa & Kimmerle, 2024; BaHammam, 2025; see also, Baron, 2026). When Professor Choi presents her work, however, she demonstrates expertise as she takes risks (her ideas are "spicy," she says, in her TED Talk), links ideas, and conveys confidence with her voice and gestures (Choi, 2023). This authentic, embodied expertise builds trust with the public in ways that generated text cannot. Only a human can deliver a responsive, engaging presentation that reads an audience and adjusts in real time.

A PROCESS FOR DEVELOPING AN ORAL PRESENTATION

Creating an effective oral presentation takes a number of steps, which may not happen as linearly as our numbering suggests. You might find it helpful to skip a step or to return to a step when you have greater clarity about your purpose or point. While you cannot always anticipate to the same degree what you will need to say in an impromptu speech such as that required in a media interview, many of these steps can still be useful to prepare for such an eventuality. Moreover, these steps can help you learn the science by putting it into your own words.

1. When you're preparing for an oral presentation, begin with AGE: Audience, Goal, and Environment (Oral Communication Program, n.d.).

   Audience: Spend some time reflecting on your audience, even taking notes to help you name what you expect from them. Who are they? What do they already know? What do they need to know? And what do they care about? Inclusive oral science communication will also take into account the audience's culture and accessibility needs (return to Chapter 3 for more on the latter).

   Goal: What do you want to accomplish? Are you sharing preliminary findings to get feedback? Explaining established research to build broader understanding? Demonstrating a new technique? Or publicizing the implications of recent findings?

   Environment: What kind of space will you be speaking in? Will you have reliable technology? What comes before and after your talk? And perhaps most important, how much time do you have? Understanding these constraints helps you prepare appropriately.

2. *Identify Your Takeaways*. Before drafting slides or outlining, identify the one to three main points you want your audience to remember. If they forget everything else, what must they retain? Write these points down as simple sentences. This step will help you *distill* your content for a listening audience.

3. *Build Your Narrative Structure*. When you're communicating science in public, a story is a great way to capture and keep the audience's attention. Map out the stasis-disruption-resolution arc described above. What's the settled understanding? What problem or question disrupts it? How does the science you've collected address it? This gives you a skeleton for the full presentation. Your presentation might also follow a character who changes over time; for more about this story technique, see Chapter 2 on rhetorical tools or Chapter 7 on profiles.

4. *Select Your Best Evidence*. You can't include everything from your written work or everything that you've learned. Choose the examples, data, or analysis that most powerfully support your core messages. Two strong examples beat five mediocre ones.

5. *Draft in Oral Style*. Write your talk in language that's easy to hear. Read sentences aloud as you draft them. If you stumble over a phrase, rewrite it. Aim for shorter sentences, active voice, and concrete language. Define technical terms immediately when you introduce them.

6. *If Tech Permits, Design a Slide Deck*. Some people find it easier to think through slides than in a script. Steps 3–6 are often accomplished in slideware such as Microsoft PowerPoint or Google Slides. If you're struggling to identify your talk's key message or structure, try making slides to help you invent and find the best order of ideas (it's easy to move slides around in a deck). Then edit ruthlessly. Each slide should have a clear purpose—showing data, illustrating a concept, or providing a memorable image—with nothing extraneous.

7. *Get feedback*. Present to colleagues, friends, or family—anyone who can tell you where they got confused or lost interest. Ask specific questions: Was my main argument clear? Did the third example help or seem redundant? Were my slides readable?

8. *Practice Out Loud, Repeatedly.* Speak through your presentation until you're comfortable delivering it. Notice where you stumble over words, where the logic feels unclear, and where you yourself feel your momentum lagging. Revise accordingly. Time yourself every time you practice and adjust content to fit your constraints. Going over time is distressingly common and disrespectful to both audiences and other speakers. If you must cut content, trim examples or details rather than eliminating entire arguments.

   Once your content is solid, focus on how you deliver it. We strongly recommend that you video yourself and then watch it back. This can be hard to do, so go easy on yourself! Still, you want to assess your speaking as dispassionately as you can. What strengths can you build on? What requires more attention and practice? Are you varying your vocal pace, tone, and volume? Using gestures naturally? Projecting confidence?

9. *Prepare for Questions.* Anticipate what you might be asked and consider how you'll respond. For example, are there policy implications to the science that many may care about but you haven't had time to address? Consider sketching those out before you speak. Sometimes it can be helpful to have a "back-pocket" story or project in mind to talk about, too. You might also seed questions that you'd like to answer throughout your presentation. For example, "I don't have time to say more about this methodology now but would be happy to take questions about it in the Q&A if anyone is interested."

10. *Embrace Revision.* Your first draft will likely be incomplete and unfocused. Strong presenters revise extensively based on rehearsal and feedback, learning about their content, the best ways to explain it, and their strengths as oral communicators in every practice session.

TAKEAWAYS

- Oral communication is integral to a number of genres, each of which requires different strategies.
- Effective oral communication helps listeners process and retain information.

- Visual support through slides should reinforce rather than replace your spoken words
- Extemporaneous delivery—carefully prepared but not scripted—allows you to speak naturally while ensuring you cover essential content.
- Practicing oral communication strengthens your thinking and makes you a more effective researcher overall.

FURTHER RESOURCES

Bugstahler, S. (2022). Equal access: Universal design of your presentation. University of Washington. https://accesscomputing.uw.edu/brief/equal-access-universal-design-of-your-presentation/

Duarte, N. (2008). *Slide:ology: The art and science of creating great presentations.* O'Reilly Media.

Reynolds, G. (2008). *Presentation Zen: Simple ideas on presentation design and delivery.* New Riders.

"TED Science Standards." TED conferences. https://www.ted.com/about/our-organization/our-policies-terms/ted-science-standards

REFERENCES

Abrahams, M. (2016). *Speaking up without freaking out: 50 techniques for confident and compelling presenting* (3rd ed.). Kendall Hunt Publishing.

Alder, K. (2012). Scientific conventions: International assemblies and technical standards from the republic of letters to global science. In M. Biagioli and J. Riskin (Eds.), *Nature engaged: Science in practice from the renaissance to the present* (pp. 19–39). Palgrave Macmillan.

BaHammam, A. S. (2025). The transparency paradox: Why researchers avoid disclosing AI assistance in scientific writing. *Nature and Science of Sleep, 17,* 2569–2574. https://doi.org/10.2147/NSS.S568375

Baron, N. S. (2026). *Reader bot: What happens when AI reads and why it matters.* Stanford University Press.

Booth, W. C., Colomb, G., Williams, J. M.., Bizup, J., & Fitzgerald, W. T. (2024). *The craft of research* (5th ed.). University of Chicago Press.

Carson, Rachel. (1962/2012) *Silent spring.* Penguin.

Cazden, C., Cope, B., Fairclough, N., Gee, J., Kalantzis, M., Kress, G., Luke, A., Luke, C. Michaels, S., & Nakata, M. (1996). A pedagogy of multiliteracies: Designing social futures. *Harvard Educational Review, 66*(1), 60–92.

Choi, Y. (2023, April). Why AI is incredibly smart and shockingly stupid [Video]. TED Conferences. https://www.ted.com/talks/yejin_choi_why_ai_is_incredibly_smart_and_shockingly_stupid

Elbow, P. (2011). *Vernacular eloquence: What speech can bring to writing*. Oxford University Press.

Gernsbacher, M. A. (2015). Video captions benefit everyone. *Policy Insights from the Behavioral and Brain Sciences*, *2*(1), 195–202. https://doi.org/10.1177/2372732215602130

Harris, A. (2019). *Conscious: A brief guide to the fundamental mystery of the mind*. Harper Collins.

Harris, A. (2025). *Lights on: How understanding consciousness helps us understand the universe*. Audio Documentary. Macmillan Audio.

Horner, W. B. (1988). *Rhetoric in the classical tradition*. St. Martin's Press.

Johnstone, K. (1981/1992). *Impro: Improvisation and the theater*. Routledge.

Koegel, T. J. (2007). *The exceptional presenter: A proven formula to open up and own the room*. Greenleaf Book Group Press.

Lermann Henestrosa, A., & Kimmerle, J. (2024). The effects of assumed AI vs. human authorship on the perception of a GPT-generated text. *Journalism and Media*, *5*(3), 1085–1097. https://doi.org/10.3390/journalmedia5030069

Oral Communication Program. (n.d.). *An overview of effective speaking* [handout]. Stanford University.

Ted Conference. (n.d.). LLC. "About." https://www.ted.com/about. Accessed 6 Mar. 2026.

Vanderplank, R. (1988). The value of teletext subtitles in language learning. *ELT Journal*, *42*(4), 272–281. https://doi.org/10.1093/elt/42.4.272

CHAPTER 10

Data Visualization

Communicating Science for the Eye

WHAT YOU'LL LEARN IN THIS CHAPTER

- Why data visualization matters for public understanding of science
- A brief overview of historic milestones in data visualization
- An understanding of how modern software and AI are impacting data visualization
- Core principles of good chart design and chart types for science communication
- A process for creating your own charts and graphs
- A modeled example of how to move from sketches to polished data visualizations
- Practice activities to improve your data visualization skills

INTRODUCTION

In today's data-saturated world, a single well-designed chart can often reach more people and communicate information more quickly than an academic article. Although poorly designed visuals that rely on incorrect chart types, distorted scales, and "chart junk" can easily mislead or overwhelm (Cairo, 2013; Tufte, 2001), when done well, visualizations compress

DOI: 10.1201/9781003520887-12

complex evidence into patterns the brain recognizes *at a glance*, allowing audiences to draw clear insight and even take action. In this way, data visualizations, like the other genres in this book, are more than the visual display of information; indeed, data visualizations are arguments and their data are the premises upon which scientists and the public make inferences. While scientists often prize an ethos of "objectivity," simply showing raw data rarely works for non-experts who lack the background to interpret it. Instead, lay audiences need clear, explanatory charts to help them infer meaning.

This chapter will show you that making effective explanatory charts for public communication is a learnable skill. We will move from history and principles to a hands-on process for turning numbers into compelling stories that also respect the integrity of the data. Use this toolkit to complement the other genres in this book—whether for a podcast, poster, research talk, or a graphic for an op-ed—to increase the comprehension and impact of your work.

A BRIEF HISTORY OF DATA VISUALIZATION

Throughout history, data visualization has evolved alongside human culture. From early numeric records to modern interactive graphics, visual data representation has shaped how humans observe, interpret, and communicate complex information. Earlier moments in the timeline reflect both where humans were evolutionarily and the ways in which they met the needs of those moments with creative ways of visual thinking, using contemporary materials and technologies to make meaning of the world and their place in it.

Early Civilizations: Recording, Ordering, and Mapping the World

Our earliest forms of data visualization were likely data tables. Extant records of 5,000-year-old Sumerian clay tablets from 3000 BCE show the use of tablets to record grain rations, livestock counts, and labor allocations in tabular form, laying the groundwork for human cultures to organize data storage and retrieval. This function is the cornerstone of modern record-keeping and serves as a powerful reminder that even the humble table is a powerful form of visualizing data.

Some thousand years later, around 2000 BCE, ancient Babylonian, Egyptian, and Chinese astronomers began making sense of our place in relation to the cosmos, creating some of the earliest spatial data visualizations in the form of star maps. As ancient cultures expanded their empires and sought resources, mapping the sky laid the foundation for conceptualizing how to visualize the land through systematized representations of terrain, borders, and topography to aid in navigation and planning. Earliest records credit Eratosthenes (third century BCE) with developing the notion of latitude and longitude, elaborated centuries later by Claudius Ptolemy, whose *Geographie* (150 CE) systematized the concept into a formal coordinate grid for mapping the globe and inspired global cartography for centuries to come (Dilke, 1987).

Medieval and Early Modern Europe: Visualizing Change and Comparison

Upon learning how to map our sense of space, humans then began to explore ways to visualize and understand time. Jumping forward to the Middle Ages, we find early mechanical devices and manuscripts that began illustrating *movement in time*. French mathematician Nicole Oresme is credited with being among the first thinkers to solve the problem of discrete quantitative comparison (Beninger & Robyn, 1978). His *Latitude of Forms* (1505 CE) used geometric diagrams to represent how qualities like speed changed over time, anticipating the basic structure of modern bar charts. Figure 10.1 is an illustration from Oresme's *Latitude of Forms* (1505 CE) of what are believed to be some of the first bar charts in modern history. Building on Oresme's work some 300 years later, René Descartes' *La Géométrie* (1637 CE) formalized the two-dimensional coordinate plane, making it possible to plot quantitative data with precision. Together, the visual reasoning of Oresme and Descartes helped shape the foundations of early statistical visualization.

FIGURE 10.1 Oresme's bar charts, pictured in the left column, are believed to be some of the first in history. (Adapted from Oresme, 1505).

The Enlightenment: Visualizing Economy and Governance

Building on this conceptual foundation, the Enlightenment marked a turning point in how data visualization was used not only to model abstract quantitative relationships, but also to track real-world economic and political change. During this period, analysts began plotting population, trade, and financial data to reveal patterns in commerce and governance. William Playfair's *The Commercial and Political Atlas* (1786), a key precursor to modern business and economic visualizations, popularized line charts to display change over time. His visualizations demonstrated how time-series and proportional data could be interpreted clearly and quickly, establishing charts as tools not just for record-keeping, but for economic reasoning and political argument. Figure 10.2 is a time-series line chart from Playfair's *Atlas*. Like many of the charts in the book, it tracks the trade balances between England and other countries over the span of 50 or more years. Through strategic use of annotation and color—note the well-labeled, pink shaded ribbon running through the center of the graph, emphasizing the trade imbalance *against* England's economic favor—Playfair clearly focuses the viewer's attention on the chart's main idea, creating powerful arguments for England's trade policies with foreign countries.

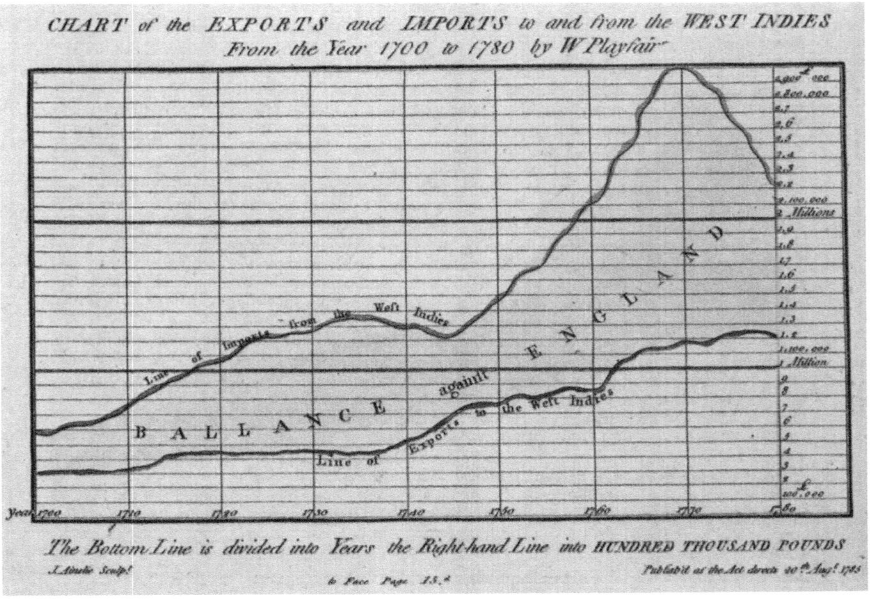

FIGURE 10.2 Exports and imports to and from the East Indies. (Reprinted from Playfair's *The commercial and political atlas*, 1786.)

The Nineteenth Century: Visualizing Disease and Death

In the nineteenth century, as industrialization fueled population booms, the parallel crises of urban overcrowding and war transformed data visualization from an economic tool into a moral necessity. This shift offered a new way to communicate and grapple with mass death, turning large-scale human loss into something visible, intelligible, and actionable. A landmark of this shift is John Snow's 1854 cholera dot map, illustrated in Figure 10.3, which utilized spatial data to visualize death in the service of public health reform. By plotting individual fatalities on the street map of London's Soho district, Snow revealed a striking cluster around the Broad Street water pump, providing him the visual evidence needed to identify contaminated water as the source of the outbreak and to overturn prevailing theories of cholera as an airborne disease.

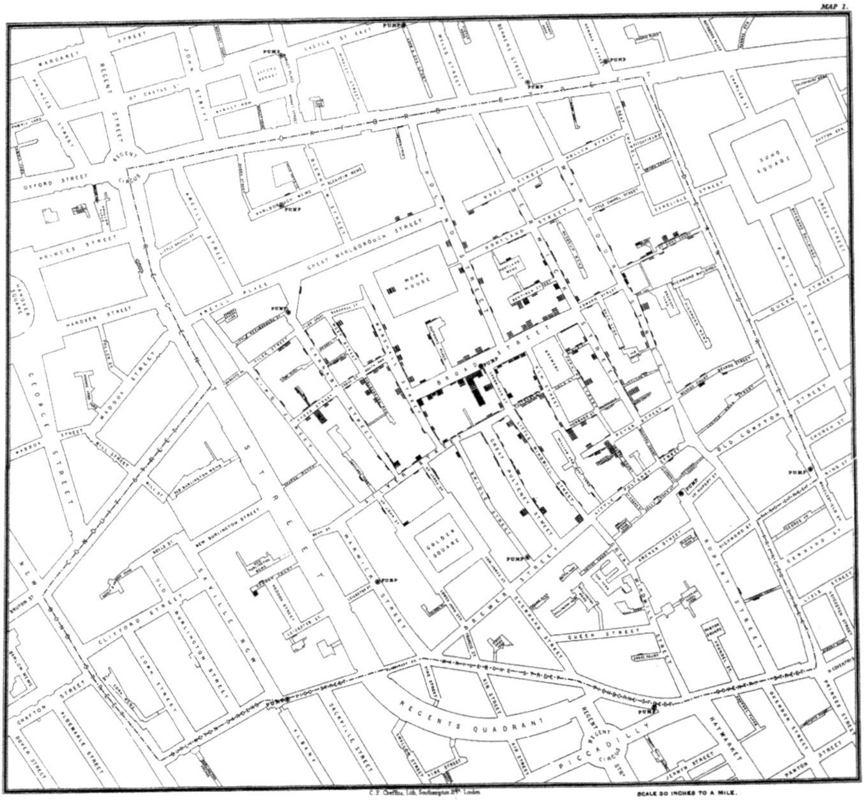

FIGURE 10.3 John Snow's Broad Street cholera map (1854). (Adapted from Snow, 1854.)

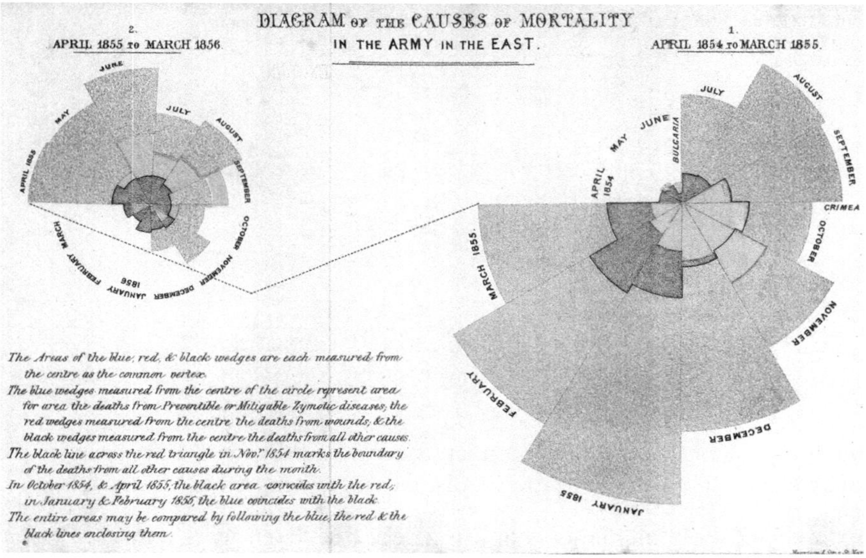

FIGURE 10.4 Nightingale's rose chart changed sanitation practices and elevated nursing to a professional practice. In the first panel in the bottom right, the larger rose, covering the period from April 1854 to March 1855, shows that deaths from disease dominate the graphic, while the smaller upper left panel depicts deaths the following year, revealing a dramatic reduction in disease-related mortality after sanitary reforms were implemented. (Adapted from Nightingale, 1858.)

A few years later, the Crimean War provided the backdrop for Florence Nightingale to advance this use of visualization to support public health advocacy. As shown in Figure 10.4, her iconic Diagram of the Causes of Mortality in the Army in the East utilize two distinct rose chart diagrams to show mortality rates before and after sanitary reforms. In both charts, the area of each wedge corresponds to the number of deaths. To interpret this, compare the large diagram on the right (April 1854–March 1855) with the smaller diagram on the left (April 1855–March 1856). In both, the large light-blue outer wedges representing deaths by preventable diseases significantly dwarf the darker, smaller wedges closer to the center, which represent death by battle wounds and other causes. The visual evidence powerfully demonstrated that even as the war continued, the primary and tragic cause of death—unsanitary conditions—could be drastically reduced through systemic reform.

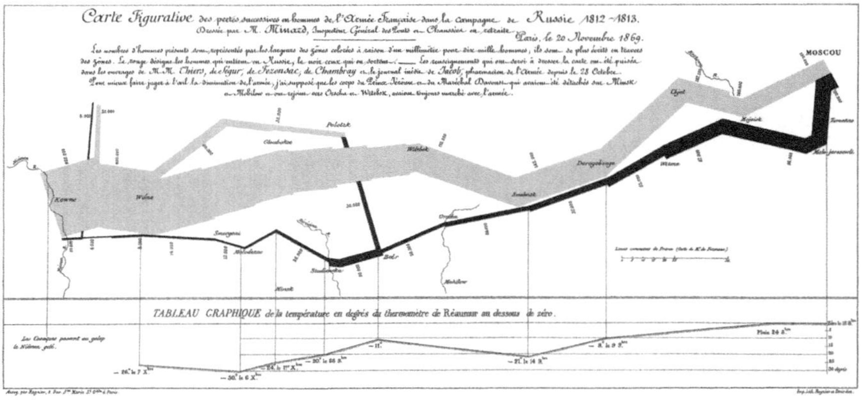

FIGURE 10.5 Minard's infamous map is considered by many to be an example of excellence in data visualization. (Adapted in grayscale from Minard, 1869.)

Like Snow's cholera map, Nightingale's rose charts were powerful visual arguments, helping to persuade British policymakers to improve sanitation and elevate public health and nursing as professional practices.

A culminating example of nineteenth-century visualization of mass death is Charles Minard's 1869 map of Napoleon's Russian campaign, shown in Figure 10.5. Unlike Snow's and Nightingale's charts, which aimed to prompt immediate policy reform, Minard's map is a retrospective visualization that memorializes a tragic past, stripping away the "glory of war" to reveal its brutal arithmetic. Consider the width of the central flow lines: the thick, light-gray band tracks the 422,000 soldiers who began the 1812 advance, while the thin black line represents the mere 10,000 soldiers who survived the retreat.

By integrating geographic movement and troop size with the sub-zero temperatures plotted along the bottom timeline, Minard made the scale of human loss immediately visible. Widely cited since its publication, Minard's multivariate map is considered by many an exemplar of storytelling with data, demonstrating how charts can move beyond recording facts to telling data-rich, analytically powerful stories.

The Twentieth Century: Theorizing How Graphics Work

While the nineteenth century proved that data visualization could save lives, the twentieth century focused on why some charts work better than others. Important at this time was statistician John Tukey's (1977), which reframed charts as tools for exploration, not just presentation, helping researchers find patterns before drawing conclusions. Jacques Bertin's *Sémiologie Graphique* (1967) was formative in codifying how visual variables—such as position, size, and color—serve as the building blocks used to encode data intuitively. But the century's most influential design philosophy arrived with Edward Tufte's *The Visual Display of Quantitative Information* (2001). In it, he argues for radical clarity, urging designers to maximize the *data-ink ratio* by stripping away *"chart junk"*—the decorative grid lines and elements that distract from the data's integrity. Together, Bertin, Tukey, and Tufte established the core principles of *exploration, encoding,* and *economy* that guide modern science communication today and which you see demonstrated in our example charts in the remainder of the chapter.

The Digital Age: Interactive, Animated, and Audience-Centered Visualization

The past two decades have transformed visualization from static images into interactive, audience-driven experiences. Advances in software and web design now allow non-experts to engage with real-time data feeds and animated graphics, shifting the focus from simply seeing results to exploring data and uncovering patterns firsthand. A landmark example of this is Hans Rosling's *Gapminder* project. In Rosling's (2006) TED Talk sharing the results of his work, Rosling used animated bubble charts to compress time, GDP, and life expectancy into a single, powerful narrative about a world that has largely moved beyond the mass death and systemic crises of the nineteenth century toward a more prosperous, longer-living global society. His work echoes the graphical excellence of Minard, proving that dynamic visuals can make global development patterns—often buried in static tables—immediately legible and powerfully persuasive.

This audience-centered approach has since expanded into public education and journalism. Projects like the American Museum of Natural History's *Human Population Through Time* (2023) map 200,000 years of change in minutes, while interactive dashboards from Johns Hopkins University and *Our World in Data* allow millions to track global challenges like COVID-19 or climate change in real time. These tools demonstrate how effective visualization now functions as a vital, transparent resource for public policy and global understanding.

Across centuries, effective visualizations share a common goal: they surface insight more quickly and clearly than text alone. Each historical moment highlighted in this chapter introduced new tools and expectations for how scientific and statistical information could—and should—be communicated visually. Today, the challenge of data visualization is no longer a lack of data or technology, but how to combine clarity, accuracy, and engagement to communicate complex information responsibly to diverse audiences. The remainder of this chapter turns from history to practice, outlining core principles that can help you meet that challenge in your own work and demonstrating a clear process for developing better data visualizations on your own.

TECHNOLOGY, AUTOMATION, AND ETHICAL RESPONSIBILITY IN DATA VISUALIZATION

Data visualization has always been a balance between creative insight and technical effort, and new technologies are accelerating both sides of that equation. Historically, visualization was slow and labor-intensive. For millennia, people literally carved, painted, or sketched their data by hand. Babylonian astronomers inscribed numerical tables into clay, and Renaissance cartographers painstakingly drafted sea charts. Every map, diagram, or statistical graphic required hours or weeks of meticulous manual labor. In that world, each design decision was deliberate: symbols, layout, and labels were chosen carefully, and each draft carried a real cost.

The rise of modern computing in the twentieth century dramatically altered this balance. With desktop computers came software such as SPSS, Excel, and MATLAB, compressing weeks of work into minutes. Analysts could enter data, select a chart type, and instantly produce a visualization. More flexible tools, including R and Python's visualization libraries, extended this capacity by allowing greater customization and programmatic control. John Tukey's call to use graphics as tools for exploration, not just explanation, became accessible to a far broader range of researchers.

Today, artificial intelligence is accelerating visual production even further. Software can automatically recommend chart types based on data structure, while generative dashboard tools assemble coordinated visuals and interactive filters almost instantly and in near real time. AI can draft captions, summarize trends, and generate code for complex interactive graphics in seconds. These tools significantly reduce the friction of prototyping, making visual exploration faster and more accessible than ever. This speed is a gift, but it also introduces new risks. Automated systems can hallucinate data, reinforce biased defaults, or obscure critical design decisions. An AI-generated chart may smooth data, truncate axes, or prioritize visual symmetry without signaling those choices to the user. If you do not understand how a tool selected a scale, filtered values, or encoded categories, you cannot meaningfully evaluate the result. In public-facing science communication, these failures can erode trust with stakeholders and audiences.

In an age of automation, *ethical data visualization* requires active human judgment. Scientists and communicators must verify every number, check every scale, and cite all data sources. If used, AI assistance should be disclosed, not hidden, and treated as a starting point for approaching chart design rather than an authority on it. Above all, AI should function as an assistant, not an oracle. Your domain knowledge, critical reasoning, and ethical responsibility remain essential to ensuring that visualizations are accurate, transparent, and worthy of public trust.

CORE PRINCIPLES OF EFFECTIVE DATA VISUALIZATION

If every chart is a rhetorical act—a visual conversation between you and your audience—then successful design requires more than just accurate data; it requires a clear purpose. As Scott Berinato (2016) notes, "good charts" are those that respect the data while surfacing insights with maximum efficiency. To achieve this, we follow five core principles.

Principle 1: Know Your Audience and Your Purpose

Effective visualization is never "one size fits all." It begins by answering two critical questions: *Who is looking at this, and what do they need to do with the information?* A chart that succeeds for a peer-reviewed journal might fail entirely on a social media feed. Most scientific communication targets three broad audience types:

Expert Audiences: Researchers and Scientists

These readers are trained to navigate dense information. They use charts as tools to test hypotheses, compare models, or support technical decisions. They expect to see statistical notation, measures of uncertainty, and multivariable displays.

Design Tip: Prioritize precision and analytical depth over simplicity. Exploratory formats that allow the viewer to "dig in" are appropriate here.

Informed Professionals: Policymakers, Educators, and Science Communicators

These readers understand the broader field but may not be specialists in your specific dataset. They aren't looking to "explore" the data themselves; they need to understand its *implications* to make decisions, draft legislation, or teach others. They value accuracy, but they prioritize *relevance.*

Design Tip: Balance technical accuracy with clarity. Use "explanatory" formats that highlight specific trends and provide clear signposting to guide the reader to the "so what?" of the data.

The General Public: Non-Specialists and General Interest Readers

This is often the most challenging audience. They likely have no prior knowledge of your jargon and have a limited budget for "mental effort"—meaning they will move on if a chart feels confusing or requires "work." For this group, successful charts often tell a *singular, compelling story or insight* that connects to their daily lives or values.

Design Tip: Prioritize accessibility and narrative. Avoid jargon, use intuitive encodings—such as the size of a bubble to represent the size of a population—and ensure the most important insight is the most visually dominant element. Simplicity is not about "dumbing down" the data; it's about removing distraction.

Once you've identified your audience, you must clarify the specific purpose of your visualization. Cole Nussbaumer Knaflic (2015) refers to this as the **Big Idea**: a single sentence that captures not only *what* the data shows, but *why* it matters to your *specific audience.*

Think of the Big Idea as your "design north star." It acts as a filtering tool during the creation process, helping you decide what to include, what to emphasize, and what to leave out. If a piece of information doesn't directly support your Big Idea, it is likely just "noise" that will distract from your message.

Principle 2: Choose the Right Chart Type

Our second core principle is to match the chart type to your data and your purpose. Data visualization is not a "one-size-fits-all" task. You cannot simply plug your numbers into any random template and expect your audience to reach the right conclusion; the specific chart type you choose fundamentally changes how a viewer interprets your data.

To pick the right one, you must first decide which "mode" of visualization you are in:

Exploratory Charts (The "Discovery" Phase): These are internal visualizations used during the data analysis phase to investigate patterns, test hypotheses, and "herding cats." They are often dense and unpolished because their primary audience is *you*, the analyst.

Explanatory Charts (The "Communication" Phase): Once you have discovered your "Big Idea," your role shifts to the "last mile" of data work. Your goal is to translate those findings into a polished form that a broader, generally more lay audience can grasp *at a glance.*

Choosing the right chart type is therefore not a matter of aesthetics, but of effective communication with the right audience. The focus of this chapter is on *Explanatory Charts,* where design is not about aesthetics, but about matching *function to form.* Different *types* of data call for different types of charts.

Step 1: Identify the Function

Choosing the wrong chart type can confuse or mislead. The first step in deciding your chart type is to determine what your data is trying to show. Before picking a chart "genre" (like a bar or line chart), ask: *What is my visual objective?* Most explanatory charts serve one of five primary functions:

Comparison: Do you want to compare values across different categories? (e.g., *Average exam scores by department*)

Change over Time: Do you want to show a trend or a trajectory over a chronological period? (e.g., *Global temperature shifts across decades*)

Distribution: Do you want to reveal how values are spread or clustered within a single dataset? (e.g., *The range of student ages in a specific classroom*)

Relationship: Do you want to show how two or more variables interact or correlate? (e.g., *The link between hours studied and exam performance*)

Composition: Do you want to show how individual parts contribute to a whole? (e.g., *A university's budget breakdown where values total 100%*)

As we stated in Principle 1, identifying your chart's *purpose* first makes chart selection more intuitive. Once your purpose is clear, the chart type usually follows naturally.

Step 2: Choose the Chart Type

Once you have identified your chart's function, the "genre"—the specific visual form such as a bar, line, or scatter plot—usually follows naturally. Choosing the right genre reduces *cognitive friction,* allowing viewers to focus on your takeaway rather than struggling to decode the visualization.

When selecting a genre, consider these standards in scientific communication:

- *For Comparison*: Start with *bar charts*; they are the gold standard for allowing the eye to compare precise differences in magnitude across categories.
- *For Change over Time*: Start with *line charts*; they intuitively imply a continuous sequence and are ideal for tracking trends or trajectories.
- *For Distribution*: Start with *histograms or box plots*; these are designed to reveal the "shape" of your data, including clusters, gaps, and outliers.
- *For Relationship*: Start with *scatter plots*; they are the most effective tool for visualizing how two or more variables interact or correlate.
- *For Composition*: Start with *stacked bars*, or use *pie charts* sparingly; these work best when categories are few and the goal is to emphasize proportion rather than precise comparison.

It is worth noting that charts showing composition should be used sparingly. Pie, donut, and stacked bar charts work best when the number of categories is small and the goal is to emphasize proportion rather than precise comparison. When exact differences matter, bar charts are often more effective than a pie or donut chart.

WHICH CHART TYPE SHOULD I USE?

If Your Question Is...	You Are Trying to Show...	Start with ...
Which group is higher or lower?	Comparison across categories	Bar chart
How does this change over time?	Trends or trajectories	Line chart
How are values spread out?	Distribution and variation	Histogram or box plot
Are two variables related?	Association or correlation	Scatter plot
How do parts make up a whole?	Composition (100%)	Pie, donut, or stacked bar

Choosing the right chart type helps your audience understand your point quickly by reducing cognitive friction. When a chart's form clearly matches the task—comparing values, tracking change, or revealing relationships—viewers can focus on the takeaway rather than figuring out how to read the visualization. For more information about selecting chart types, refer to the Common Chart Types in the Science Communication Glossary for visual examples and specific use cases in scientific research.

Principle 3: Prioritize Graphical Integrity

Bad charts are everywhere, from social media and news outlets to respected academic journals. While some are merely cluttered, the most dangerous charts are those that mislead—whether intentionally or not (Cairo, 2019). In rhetorical terms, this principle is about *ethos*: the credibility and trustworthiness of the communicator. If your audience detects even one "trick" in your data, they will likely discard your entire argument.

In science communication, your visual must be a faithful representation of the underlying data. Honesty is not just the absence of "fake data"; it is the commitment to ensuring that the *visual magnitude* perceived by the viewer matches the *numerical magnitude* of the evidence. To maintain this integrity, you must apply different rules depending on whether you are showing amounts or trends. Here are three rules to follow to maintain graphical integrity as you build your visualizations:

Show the Full Context (Avoid Cherry-Picking)

Honesty requires context. While your chart's final design will work toward "focusing the eye" on a specific insight, you must never do so by suppressing data that complicates your narrative. A chart with high graphical integrity presents enough data to allow the viewer to see the "larger trend" rather than a convenient snapshot. Here are three tips to keep in mind:

Provide a Sufficient Timeframe: Avoid selecting a narrow window of time just because it supports a specific claim. If a 5-year trend shows growth but the 20-year trend shows a steady decline, showing only the 5-year "slice" is a visual half-truth.

Acknowledge Outliers: Do not omit outliers simply to make your chart look cleaner or your trend line look more significant. In science, outliers are often the most important data points; they indicate anomalies, errors, or new phenomena that require explanation.

Be Transparent About Data Gaps: If data is missing for a specific period or variable, do not "connect the dots" across the gap with a solid line. Instead, use a broken line or a visual marker to indicate the missing evidence.

Respect the Baseline (Showing "How Much")

The baseline is the starting point of your scale. Because our brains perceive the value of a graphic by its physical size or distance from this starting point, where you begin your axis changes the truth of the chart.

For Bar Charts: Because a bar represents a quantity through its length, the axis *must* start at zero. Truncating the axis (e.g., starting at 50 instead of 0) visually exaggerates small differences, making a 5% increase look like a 500% jump. This is the most common way data is accidentally "weaponized" to mislead.

Consider: In a bar chart of annual CO_2 emissions, starting the y-axis at 35 gigatons instead of 0 would make a 1-gigaton increase look like it tripled the total global output.

For Line Charts: Line charts track *rate of change* (slope) rather than quantity. In some scientific contexts, starting at zero can "flatten" a line and hide critical fluctuations.

Consider: When plotting global surface temperature anomalies, the difference between +1.0°C and +1.5°C is severe. Starting the y-axis at 0°C (or absolute zero) would render this vital increase as a flat, seemingly meaningless line at the top of the chart. In this case, "zooming in" on the relevant range is the more honest choice, provided the axis is clearly labeled.

Keep Scales Consistent (Showing "How Fast")

Scale consistency ensures that the "rhythm" of your data is not distorted. If the physical distance between points on your chart does not match the numerical distance between the data points, you are distorting the "speed" of the science.

Equal Intervals: Every "inch" on your axis must represent the same numerical value.

Consider: If your x-axis plots years but jumps from 2010 to 2015 and then to 2016 in the same physical span, you are "scrunching" time. This makes a slow 5-year change look just as sudden as a 1-year change, hiding the true rate of change.

Linear vs. Logarithmic Scales: A linear scale (1, 2, 3, 4) is a good default because it matches our natural perception of growth. However, if your data spans several orders of magnitude (like viral load or earthquake intensity), a logarithmic scale (1, 10, 100, 1000) may be necessary to keep the charting of the data within a reasonable frame.

Consider: If you use a log scale to prevent your data from "shooting off the top of the page," you must explicitly label it as such. A reader who assumes a log scale is linear will fundamentally misunderstand the scale of the phenomenon.

Maintain Proportionality (Area and Volume)

When using shapes like bubbles or icons (pictograms) to represent data, the *area* must be proportional to the value, not the height or radius.

Consider: If you double the radius of a circle to show a "doubled" value, you actually *quadruple* the area (since $Area = \pi r^2$). To the viewer's eye, the data now looks four times larger than it actually is. Always scale by area to avoid visually overstating your findings.

Principle 4: Declutter Ruthlessly

Once you know your audience and big idea, have selected an appropriate chart type, and established a scale or proportion that maintains graphical integrity, your next goal is to make the data easy to read. In scientific communication, clarity is not just an aesthetic choice—it is a functional one. A chart can be technically "honest" but still fail if the viewer cannot distinguish the signal from the noise.

Effective visualization often requires multiple rounds of revision. Building on Edward Tufte's (2001) philosophy, the most powerful way to refine a chart is through *strategic subtraction*. By identifying and removing *chartjunk*—unnecessary 3D effects, decorative icons, and heavy gridlines—you reduce visual "noise" and increase the *data-ink ratio.* Every element that remains on the page should contribute to the viewer's understanding; none should be merely decorative.

While some research suggests that icons or pictograms can aid recall for general audiences (Few, 2011), these elements frequently read as *clutter.* Use them sparingly and only when they serve a specific communicative intention (Berinato, 2016). To streamline your visualization and prepare the viewer's eye for the "Big Idea," focus on decluttering these two high-impact areas:

Legends: If your chart only has two or three categories, label the data points directly. This eliminates "split attention," where a viewer must constantly look back and forth between the legend and the data.

Gridlines and Axes: Gridlines often clutter the plot area. Remove them entirely if they aren't essential, or lighten them significantly so they stay in the background. If a specific data point is critical, label it directly rather than forcing the reader to squint at an axis.

> **Design Tip:** Every mark on the page must "earn its place." Ask yourself: Does this element help the viewer understand the "Big Idea" more quickly? If the answer is no, you can probably cut it.

Figure 10.6 illustrates the impact decluttering can have on visual clarity.

Using the same hypothetical dataset concerning projected global temperature changes (°C) from 2025–2100 under three different policy scenarios, note how with a few simple decluttering moves, we can reduce *cognitive load* and sharpen the "Big Idea":

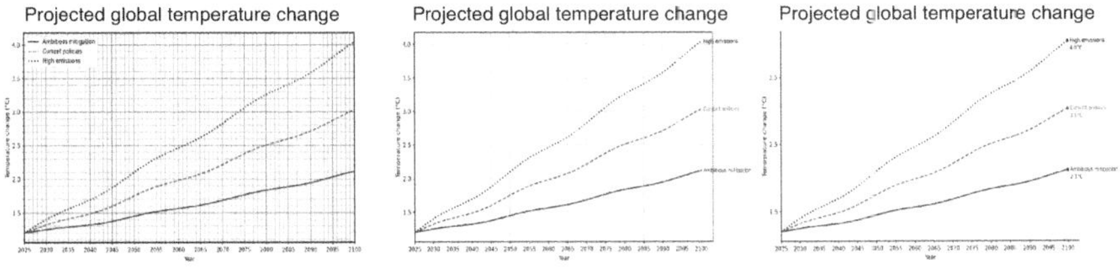

FIGURE 10.6 (A–C) How to declutter to reduce cognitive load and enhance clarity. (*Note*: All three charts display the same synthetic data; improvements in clarity result solely from changes in design).

Chart A (High Noise): This original prototype chart looks like something software would render quickly, but it lacks refinement. It uses a legend that forces the viewer to shift attention back and forth to identify the lines. Heavy gridlines and dense tick marks add visual noise that competes with the data, slowing down comprehension and lowering the data-ink ratio.

*Chart B (Improved Focus)***:** By ***labeling the lines directly***, we eliminate the legend and allow the viewer's eye to stay on the data. However, persistent—albeit fewer—gridlines still create unnecessary clutter, making the chart feel "busy."

Chart C (*Cleanest Design*): This version achieves the highest ***data-ink ratio***. We've stripped away the gridlines and legend entirely, opting for direct labels and final-value annotations.

Figure 10.6 demonstrates how effective decluttering sharpens a chart's message without sacrificing precision. By treating revision as subtraction, the final chart transforms a standard technical plot into a clear, persuasive narrative about which policies might work better than others. Every remaining mark now serves a specific purpose, allowing the audience to quickly and clearly grasp the Big Idea.

Principle 5: Focus the Eye and Establish Visual Hierarchy

If decluttering is about subtracting from the visual field, focusing is about *strategic addition* to introduce visual hierarchy. Where decluttering removes the noise, focusing adds visual emphasis that ensures the viewer knows exactly where to look first to find your Big Idea.

Color

Color is your most powerful tool for focusing because it is a *preattentive attribute*—our visual system processes differences in hue and intensity instantly, before we even consciously "read" the data. To use color intentionally, move beyond decoration and follow these core strategies:

Practice Restraint: Too many competing hues dilute emphasis. Use a neutral palette (like light grays) for background data and reserve one bold, saturated color for your focal data.

Respect Convention: Play to your audience's existing "mental models." Our brains instinctively associate red with "hot" or "danger" and blue with "cool" or "safety." Swapping these—using blue for heat and red for cold—creates cognitive friction that forces the viewer to work harder to understand your chart.

Leverage Intensity: Darker or more saturated colors naturally draw the eye, while lighter, desaturated tones recede. Use this to "push" supporting data into the background.

Prioritize Accessibility: Roughly 8% of men have color-blindness. Avoid red-green pairings when possible and use tools like ColorBrewer or Viz Palette to ensure your chart remains legible. See Chapter 3 on *Inclusive Science Communication* for more guidance on how to prioritize accessibility.

Match Palette to Data Type:

Sequential Palette: Use one color going from light to dark when emphasizing intensity or order. The darker the color, the "more" of something there is (e.g., low-to-high income).

Diverging Palette: Use two colors meeting at a neutral midpoint when emphasizing comparison or deviation. This is best for showing data that moves away from a center point, such as global temperature anomalies (where redness indicates warming and blue indicates cooling relative to a historical average).

Categorical Palette: Use distinct colors (like orange, purple, and green) when your data groups have no mathematical relationship—for example, comparing different land-use types (Forest vs. Urban) or different experiment groups (Control vs. Treatment).

Beyond Color

While color is the "loudest" focusing tool in your kit, these structural properties also provide visual cues for your viewer's eye. By strategically varying these properties, you can guide a reader through a complex dataset without them even realizing they are being led:

Size and Weight (Visual Volume): Larger and bolder elements are perceived as more important. In Minard's map (Figure 10.5), the overwhelming width of the starting line compared to the thin line of the retreat tells the story of loss through pure scale.

Position and Alignment (The Path): Western audiences naturally read from top to bottom and left to right. Information placed in the top-left corner of a chart is perceived as the starting point. Furthermore, *alignment* creates a sense of order and reduces "visual noise." When elements like titles, legends, and axes aren't aligned to a common edge, the viewer's brain works harder to "scan" the page.

DESIGN TIP: MASTERING VISUAL HIERARCHY

Once your chart is decluttered, use these tactical tips to guide the viewer's eye to your Big Idea:

1. *Color Match*: Is your color choice appropriate for your data type (sequential, diverging, or categorical)?
2. *Visual Volume*: Are your primary findings bold and your other data "pushed" into the background (using light grays or transparent colors)?
3. *Invisible Grid*: Is your text (title, legend, labels) aligned to a clean, consistent edge?
4. *Accessibility Check:* Have you avoided red-green pairings and used direct annotations to help color-blind readers?
5. *The "Squint Test"*: If you squint at your chart, is the most important element still the most visible?

Principle 6: Create Context

Even the clearest charts do not speak entirely for themselves. Data shows *patterns*, but context clarifies *meaning*. Effective visualizations guide viewers toward interpretation by clarifying what matters, why it matters, and how the data should be read. Creating context helps ensure that viewers draw conclusions that are accurate, relevant, and aligned with your purpose. The following charting elements can do much of this interpretive work:

Titles: State the Big Idea

A chart's title should communicate its main takeaway, not merely describe its contents. Action-focused titles help viewers understand what they are meant to notice before they begin interpreting the visual evidence.

Descriptive (Weak): "Average Test Scores by Class"

Action-Focused (Strong): "Section B Students Scored Highest on the Final Exam"

Using "character–action duos"—pairing a clear subject (Section B) with an active verb (scored highest)—turns the title into a concise narrative. Titles should also be placed prominently at the top of the chart, generally left-justified and aligned with the axis, to establish hierarchy and guide interpretation.

Callouts and Annotations

Brief annotations can clarify trends, anomalies, or turning points that might otherwise be overlooked. Short callouts placed near key data—such as "Temperatures declined after policy change"—help viewers connect visual patterns to explanatory context.

Reference Lines: Provide Benchmarks for Comparison

Strong titles and graphics embody Tufte's principle of "compared to what?" by framing the chart as a claim rather than mere a container of data. Reference lines (such as an average, a goal, or a historical baseline) also provide a necessary visual comparison. *Example:* A chart showing current carbon emissions is more meaningful when a horizontal reference line shows the "Target Goal" for the year.

Label Exhaustively

Every graph requires clear, detailed labels for its axes, including the specific units of measurement (e.g., "Meters," "USD"). A caption beneath the chart should also provide essential meta-information, including the *data source*, to establish credibility and allow for independent verification.

Together, these contextual elements make your chart's idea more legible and trustworthy. By pairing clear visual design with thoughtful contextual cues, you help viewers understand not just *what* the data shows, but *why* it matters.

A MODEL OF EXCELLENCE

To see how these principles work in concert, we return to Charles Minard's 1869 map of Napoleon's Russian campaign (Figure 10.5). More than a historical artifact, the map is a masterclass in how multivariate data can be synthesized into a single, devastatingly clear "visual argument."

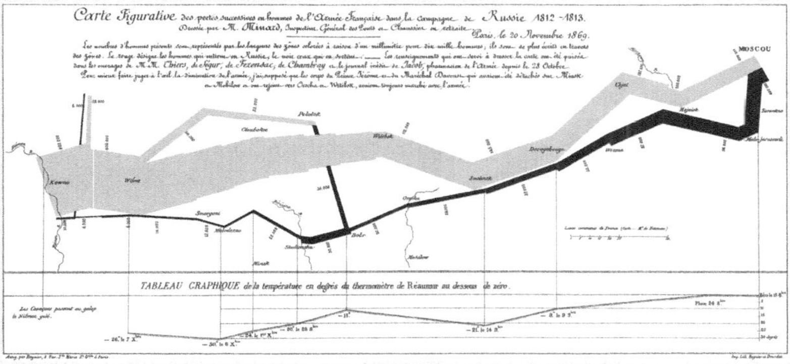

1. *Know Your Audience and* Purpose: Minard designed this for an informed public rather than for technical military planners. By balancing high density with legibility, he moved the story beyond mere "stats" into an accessible narrative of national tragedy.
2. *Prioritize Visual Integrity (Honesty)*: The map maintains statistical integrity. Troop size is encoded **proportionally through line width**, allowing the viewer to witness the gradual accumulation of loss rather than just an abstract final total. Minard builds trust through restraint; the scale itself conveys the tragedy without the need for dramatic exaggeration.
3. *Match Chart Type to Purpose*: A standard bar or line chart could not capture the multidimensional nature of this collapse. Minard's **flow map** is the perfect "genre" for his goal: it shows movement, attrition, and environmental context (temperature) simultaneously to answer a single question—*how and why did the army vanish?*
4. *Ruthlessly Declutter*: Despite its complexity, the map contains no "chartjunk." There are no ornamental borders, unnecessary icons, or decorative effects. Every mark—from the angle of the retreat to the temperature labels—carries essential information, exemplifying a high *data-to-ink ratio.*
5. *Establish Focus and Visual Hierarchy*: Minard uses *visual volume and contrast* to tell the story. The thick, light-gray band represents the 422,000 soldiers advancing; the thin, dark line represents the mere 10,000 returning. This stark contrast in "weight" pulls the eye immediately to the scale of the catastrophe.

6. *Create Context*: Minard provides the essential *"compared to what?"* by integrating a temperature chart at the bottom. By aligning the sub-zero readings with specific points on the retreat, he provides the causal context—proving that the extreme cold was as lethal a factor as the military conflict itself.

Together, these principles explain why Minard's map continues to be taught more than a century later. It shows that effective data visualization is not about flashy design or technical novelty, but about aligning purpose, integrity, form, emphasis, and context to help audiences see—and understand—complex realities. As you begin designing your own charts, Minard's work offers a reminder that clarity, honesty, and thoughtful design choices can turn data into insight.

A PROCESS FOR MAKING BETTER CHARTS

As you continue developing your data visualization literacy, your chart-making process will also continue to improve, both in terms of efficiency and execution. Having an effective process can also help. Below, we offer a five-step process that should help you make better charts that build on core principles of good chart design.

Step 1: Develop Your Big Idea

Start by clarifying your communication goal. Who is your audience? What should they learn? Why does it matter? Good charts are purpose-driven and audience-selective. Knaflic (2015) encourages people to write a one-sentence "Big Idea" using this formula:

Who needs to know + What they should learn + Why it matters

Example: "Show students (who) the strong positive correlation (what) between study time and academic performance (why)."

To expose blind spots or a lack of clarity, consider developing and getting feedback on two or three versions of your Big Idea with a partner to check for clarity and persuasiveness.

Step 2: Sketch It Out

In early-stage design, hand-sketching is your best tool for creative exploration of ways to visualize your Big Idea. Moving away from the screen and its default smart charting allows you to iterate ideas and focus on core data relationships. As Roam (2008) and Berinato (2016) emphasize, working "analog" lowers the stakes and encourages the intuitive thinking necessary to refine your message.

When considering what to sketch, try the following approaches to pressure-test your Big Idea:

Experiment with Chart Types: Swap a stacked bar for a line graph to see which emphasizes your comparison better.

Experiment with Encoding: Swap out colors or change the way you highlight variables to see how changing preattentive attributes focuses the eye differently.

Iterate Quickly: Spend 15 to 30 minutes generating two to four variations that change the design without altering the data's meaning.

Invite Feedback: Use sketches as working drafts to gut-check your logic with peers before committing to digital production.

Step 3: Talk It Out

User feedback is an essential step in your visualization process. Because visual confirmation bias can blind you to confusing layouts or misleading scales, engaging with others early during the sketching phase allows you to pressure-test your message before committing to digital production. Feedback is the bridge between what you *think* you're showing and what an audience actually *sees*. This collaborative approach follows Berinato's (2016) "Talk-Sketch-Prototype" cycle. Start with the "gut check"—are you and your peer on the same page?

> Big Idea Gut Check: Before touching software, conduct a visual critique with a peer or two. Start with a 30-second gut check: if they cannot identify your Big Idea in half a minute, your visual hierarchy needs work.

Because programs visualize data while people visualize ideas, this talk stage ensures your mental model aligns with human perception. Use your partner to help you emphasize key data, frame the narrative in the title, and focus the design by identifying chartjunk. Sometimes they might even see other ways to visualize the data. Listen to this feedback; it tells you whether there is a gap between your intent and the viewer's interpretation.

To move beyond general opinions like "I like it," we offer the following structured prompts to ensure you receive actionable feedback from your peers:

1. *The Signal vs. Noise Audit*: Ask your peer to perform a quick visual scan of your sketch and answer:

   "What is the very first thing your eyes move toward?" If it isn't your primary data point, your visual hierarchy and focusing likely need work.

   "Is there anything here you find distracting or confusing?" This helps you identify "noise"—unnecessary gridlines, redundant labels, or complex legends—that you may have become blind to during the sketching phase.

2. *The Caption Test*: Ask your peer to write a *one sentence caption for the chart based only on the visual evidence.* If their caption doesn't match your intended Big Idea, your design or your Big Idea may need work.
3. *The Expert and the Novice Check*: In science communication, you often need to balance accuracy and accessibility. Ideally, ask two different peers the following:
   - *The Expert:* Talk to a peer who knows the science: Ask: *"Is this an honest representation of the data?"* (Checks for Principle 3: Integrity).
   - *The Novice*: Talk to a peer who is unfamiliar with the topic. Ask: "*What is the main takeaway here?*" (Checks for Principle 1: Audience and Purpose).

By treating a peer's confusion as a valuable data point, you transform default charting into a dialogic act of ethical, persuasive visual communication.

Step 4: Prototype a Draft

After refining your concept through sketching and feedback, you are ready to *prototype*—shifting your visualization goal from exploration to *explanation.* Tools such as Excel or Google Sheets are suitable for standard charts. Platforms such as Tableau or R offer more capabilities for complex datasets but have a steeper learning curve. AI is not recommended for prototyping because these tools may produce inaccurate data, which goes against the principle of *graphical honesty*. Given software advancements, this stage is generally pretty fast, especially if you are relying on conventional chart types like lines and bars that most software can smart chart quickly.

It is worth noting that in some cases a digital tool may not be necessary. As illustrated in William Playfair's line chart (Figure 10.2) or Minard's map (Figure 10.5), a refined hand-drawn chart can be very effective, particularly when exploring new ways of visualizing information and data. Remember, the medium should always serve the rhetorical purpose, and the idea drives the chart design, not the software.

Refine and Polish

You now have a working prototype. Let's think of this as a first draft; like any good piece of writing, revisions are often necessary. The final step in chart design is *refinement*, where you transition from a functional prototype to a polished, professional visualization. As Berinato (2016) suggests, this is the stage of "persuasion," where you clarify the narrative by decluttering and focusing strategically: remove redundant labels, background lines, and chartjunk to ensure the data remains foregrounded; apply contrast intentionally to guide the eye toward your Big Idea; and verify that your scales are accurate and proportional to maintain ethical integrity. This includes choosing accessible color palettes—such as those readable by color-blind viewers—and using titles that are descriptive rather than just topical. For example, instead of "Average Global Temperature (1880–2020)," use more effective explanatory titling like "Average Global Temperatures Have Risen 1.1°C Since 1880," which identifies a character (temperatures), an action (have risen), and includes a magnitude (1.1°C). Remember, a well-polished chart should feel purposeful rather than decorative. By simplifying the design and citing your data sources clearly, you build the credibility necessary for your audience to grasp your central insight quickly and confidently.

A Note about Data Visualization Tools

When it comes time to move from pencil to pixel, the choices can be overwhelming. In professional science communication, rarely does a single tool do it all. Most effective visualizations are the result of software layering: using one program for computation and another to help polish the design. You might start with a statistical tool like R or Excel to ensure your data is accurately plotted, but saving these as a "picture" (like a JPG or PNG) creates a raster file made of pixels that becomes blurry when edited.

TABLE 10.1 Data Visualization Tools for a Science Communication Workflow

Workflow Phase	Software Options	Primary SciCom Strength
1. Analysis and Charting *Crunching the data*	R (ggplot2)/Python Excel/Google Sheets	**Precision:** Ensures mathematical accuracy and reproducibility for peer review. **Efficiency:** Ideal for initial data organization and rapid trend identification.
2. Prototyping *Testing the concept*	Tableau Flourish	**Interactivity:** Best for building deep, searchable databases, and public portals. **Engagement:** High-impact animations and storytelling for public outreach.
3. Polishing *Crafting the story*	Canva Illustrator/Inkscape	**Composition:** User-friendly layout for "Graphical Abstracts" and posters. **Refinement:** Precise vector control for journal-ready focus and hierarchy.

To maintain *visual honesty* and clarity, you should instead export your chart as a *vector file* (like an SVG or PDF). These formats "lock" the mathematical instructions of your chart, allowing you to move into a design suite like Canva or Adobe Illustrator for the polish phase without losing image quality. In this second layer, you can swap default fonts for legible ones, add hand-drawn annotations to explain a specific data spike, and apply a color-blind-friendly palette, transforming a raw plot into a persuasive, high-resolution narrative. Table 10.1 outlines options and explanations of various software platforms that support data visualization at different stages of the process to support your chart-making goals. While some come with steeper learning curves than others, avoid chasing trends. Instead, build fluency in a few reliable tools that fit your needs; the craft is found in your design choices, not just the software you use.

Putting It All Together

Now that we've reviewed the principles for effective and ethical data visualization and provided a clear process for making your own charts, let's put this all together with a hypothetical example. Imagine you are a research assistant working on a project examining relationships between student study habits and their academic performance. As part of this research, your team has discovered a strong positive correlation between study time and exam performance. The president of the university has learned about your findings and has asked you to present this finding to incoming freshmen at orientation.

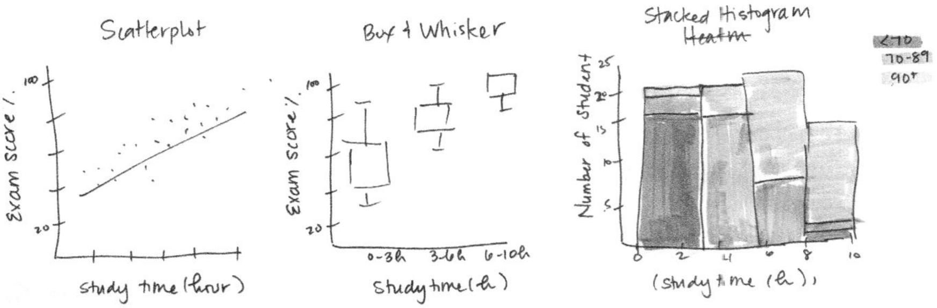

FIGURE 10.7 Visual thinking in process: Three hand-drawn sketches exploring different chart types to represent hypothetical data related to study time vs. exam scores. (Reprinted with permission from Cassie Wright)

As generalists, this audience needs to grasp your takeaway quickly to help them set better study goals. But you also have 30 minutes to talk about the research. Rather than lecturing at them, your Big Idea is to "*Show students (who) the strong positive correlation (what) between study habits and academic performance (why) as a starting point to reflect on their own study habits.*"

You begin by sketching multiple ways to visualize this trend. At this exploratory stage, the goal is to test which chart type accurately represents the data and best sparks a conversation about study habits. As shown in Figure 10.7, you might sketch three different chart types: a scatterplot to show individual data points, a box-and-whisker plot to show distributions, or a stacked histogram to show frequency. These low-fidelity sketches allow you to compare how each frame emphasizes your Big Idea before committing to a digital prototype.

With your sketches in hand, you move into the feedback stage with peers. This stage is designed to "pressure-test" which chart type best aligns your Big Idea with your audience's perception. Remember, the goal here is to move from your exploratory "internal" view of the data to an explanatory "external" one. In this simulation, your friends provide the following insights:

Friend 1 initially gravitates toward the stacked histogram, noting that the colored bins quickly highlight the high-performing group.

Friend 2 pushes back, finding the histogram and box-and-whisker plot too complex for a quick glance. They argue that the scatterplot most clearly conveys the fundamental trend, even if it feels "boring."

The Consensus: While the other charts offer more detail on frequency or uncertainty, you notice words like "complexity" applied to two of the genres, whereas your friends agree the scatterplot's simple but effective X-Y relationship clearly emphasizes the correlation between study time and grades for a generalist audience.

Given the consensus on the scatterplot as the clearest way to show a positive relationship between study time and exam scores, you open Excel, plug in the data, and start with the recommended scatterplot chart as illustrated in Figure 10.8.

The initial chart is accurate and relatively clear but visually "clunky." Heavy data points and a chaotic default trendline—which maps more like an aggregated line chart rather than a line of best fit—create unnecessary visual noise, while extraneous elements like gridlines distract from the message. This reflects a common hurdle in all data visualization software: most default "smart charts" are often the data visualization equivalent of a "shitty first draft" (Lamott, 1994). They are a necessary starting point, but they rarely constitute a finished product. To move beyond this draft, you transition from the software's defaults to a more refined, audience-centered design. Remember, you aren't just presenting data in a chart; you are refining the visual logic to ensure the Big Idea is the first and most obvious thing

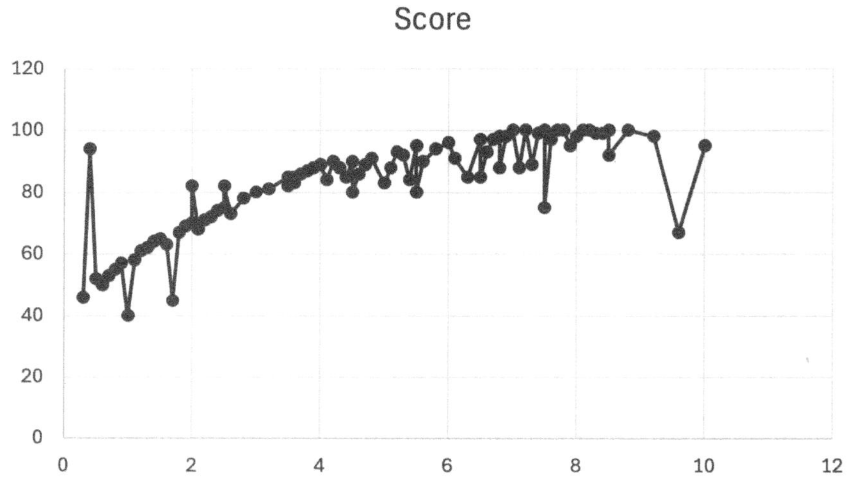

FIGURE 10.8 The SFD: a crude scatterplot prototype rendered in Excel. Data is hypothetical and used to demonstrate design principles only.

the viewer sees. Let's look at how a few strategic refinements can transform this chart from a collection of raw data into a clear and compelling narrative.

Figure 10.9 is a strategically refined chart. Removing gridlines and grayscaling individual data points declutters the visual field, allowing the viewer to focus on the most important information without distraction. Stretching the scale and emphasizing the trendline in blue provides clear visual focus, guiding the eye directly to the correlation. The explanatory title, "Predicting Excellence," gives the chart a specific purpose with character (you, the student) and action (studying), while including *r* and *n* values ensures graphical honesty by providing the context and transparency necessary for a trustworthy message. Together, these choices transform raw data into a persuasive, credible insight for the audience.

While it is tempting to think a single, well-designed chart is enough, it often isn't. This is where the final principles of sufficient context and annotation become particularly helpful. To bridge the gap between a polished chart and a persuasive argument, consider embedding your visual

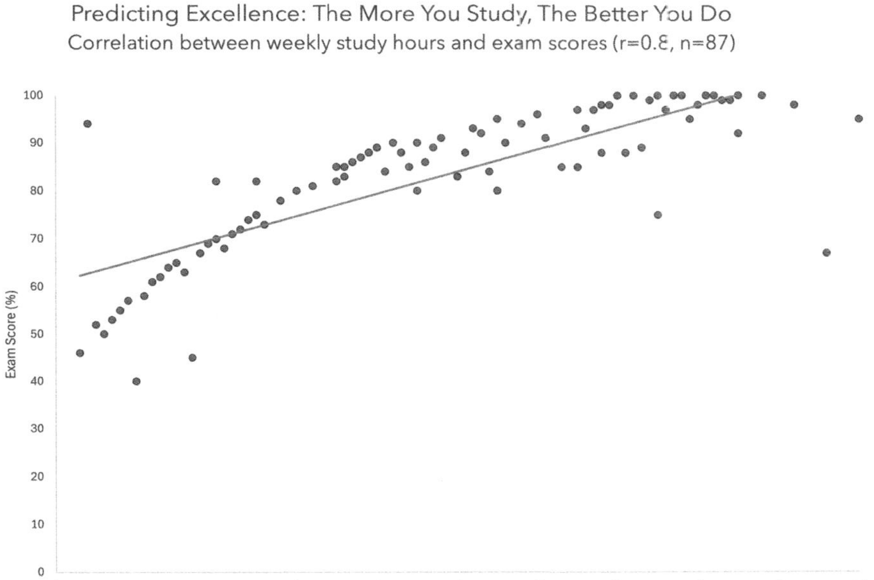

FIGURE 10.9 A polished chart. Following principles of decluttering, focusing, and contextualizing, the final chart maximizes the data-ink ratio, creating a more professional, polished chart. All changes made in Excel.

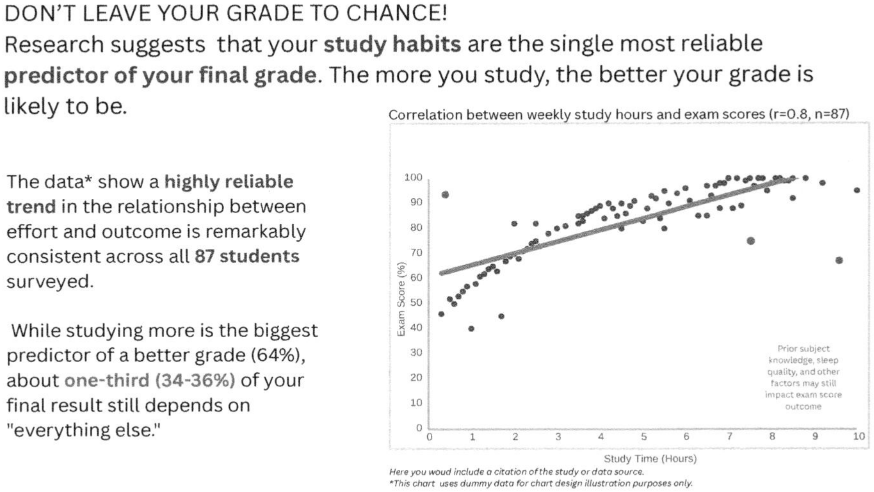

FIGURE 10.10 A data story: This final explanatory chart resembles a more clear and compelling visual narrative about the correlation between students' study habits and academic success.

within a larger narrative frame. This is as simple as building your chart into a single slide that adds layers of explanation—such as a clear Big Idea headline, highlighted data points, and concise captions—that guide your audience through the evidence. As shown in Figure 10.10, moving from a standalone chart to a fully designed slide helps ensure your message is not just seen, but clearly understood.

The slide uses a declarative headline to provide freshmen with a clear, immediate takeaway: "the more you study, the better you are likely to do" on your exams. To address the data's nuance, color-coded outliers (in red) and an annotation box in a matching color explain the interplay between random chance and study time. This is anchored by a secondary annotation that explains the study result in more detail. By adding context and annotation, and increasing the visual hierarchy with a contrasting red and blue color scheme, the final chart facilitates informed conversations about study habits and exam outcomes—the core objective of the freshman orientation presentation.

In sum, creating better charts relies on a simple process of talking, sketching, prototyping, and refining, all while keeping the core principles of data visualization (purpose, chart type, honesty, decluttering, focusing, and context) in mind. By centering the design on a single, memorable insight, stripping away nonessential elements, and adding just enough context to support interpretation, our sample charts went from crude sketches to clunky prototypes to an elegant slide that communicates its Big Idea clearly, honestly, and quickly. While the other forms—a histogram or box-and-whisker chart—could have worked, our final choice was driven by a specific purpose: guiding a generalist audience toward a clear takeaway about the benefits of studying, with some caveats. This process illustrates that there is no single "correct" way to visualize data; instead, effective visualization depends on whether your choices provide sufficient context to ensure visual clarity and honesty for your viewer. Ultimately, a "good chart" is one where every design choice is guided by these core principles to successfully communicate a Big Idea with clarity and integrity (Berinato, 2016; Knaflic, 2015).

TRY IT OUT! HANDS-ON ACTIVITIES TO IMPROVE YOUR DATA VISUALIZATION PRACTICE

ACTIVITY 1: VISCRIT

Find a chart in a news article or textbook that seems interesting but may miss the mark in some way. Explain why using the six principles of data visualization (as illustrated in our sections on Minard's map and our solar panel example). What suggestions do you have to make the chart more effective? Sketch a redesign and seek feedback on your improvements.

ACTIVITY 2: SPOT THE LIE

Find a misleading chart online and identify which principle(s) it violates. Sketch a quick correction and compare your redesign to the original. Use the six principles of visualization to justify why your design improved the clarity or honesty of the data.

ACTIVITY 3: THE PENCIL-FIRST PROTOTYPE (SKETCHING FOR INVENTION)

Using a small dataset, put away your computer and write your "Big Idea" statement at the top of a sheet of paper. Set a timer for 5 minutes and sketch three completely different ways to visualize that data (e.g., one focusing on a specific outlier, one showing a long-term trend, and one comparing categories). Reflect on which sketch felt the most rhetorically powerful and how hand-sketching changed your perspective on what was possible versus what was easy in a software program.

ACTIVITY 4: TESTING CHART TYPES

Using the same small dataset from Activity 3, quickly prototype your charts in Excel or another software of choice. Compare how they rendered digitally to your sketches. Do you still feel confident in your first choice? Which one conveys the Big Idea most effectively? Explain your reasoning based on the functional categories discussed in this chapter.

ACTIVITY 5: TESTING TOOLS

In groups, use a small dataset (either the same from Activity 3 or a new one of choice) to rapidly prototype a chart in several different software environments: Excel, Flourish, Google Charts, an AI generator, and coding software (if the ability is present). Compare the design affordances and labor efforts of each tool. Note which allowed for the greatest control, accuracy, and design flexibility.

ACTIVITY 6: KNOW THE RULES OF THE GAME

In groups, research the AI usage policies for a major journal in your field (such as *Nature*, *JAMA*, or *IEEE*) to see how they handle disclosure and visual honesty. Compare how these professional standards regulate AI-assisted charts versus "human-only" visualizations to ensure data integrity.

ACTIVITY 7: THE AUDIENCE TRANSLATION CHALLENGE

Select a single, complex scientific dataset (e.g., local air quality levels or a recent clinical trial). Create two distinct "Big Idea" statements and rough sketches for two different audiences: a specialized researcher in that field and a local community member on social media. Compare how your choice of visual volume, color intensity, and title wording shifted between the two versions.

ACTIVITY 8: MAPPING THE INVISIBLE (CONCEPTUAL VISUALIZATION)

Select a complex scientific concept that is not primarily numerical—such as the nitrogen cycle, the path of a viral infection, or the steps of a legislative bill. Identify the key "nodes" or entities in the system and define their relationships using arrows, lines, and proximity. Attempt to simplify the map so that a non-specialist can follow the logic without reading a detailed text description.

CONCLUSION

Data visualization is the art of turning numbers into clear and compelling visual stories. As this chapter has shown, charts are not neutral displays of information;theyarearguments.Theyproposepatterns,frameexplanations, and invite interpretation, discussion, and action. Across history—from early astronomical tables to Snow's cholera map, Nightingale's mortality diagrams, and Minard's account of war and winter—data visualization has played a central role in helping societies make sense of complex evidence and human consequence.

Building on this historical foundation, we examined the core principles that make visual arguments effective: clarifying a Big Idea, choosing forms that match purpose, designing honestly, reducing clutter, guiding attention, and creating context. These principles help ensure that charts do not merely look persuasive, but *earn* persuasion through accuracy, transparency, and thoughtful design. We then turned to process, emphasizing sketching, prototyping, and revision as ways to think visually before committing to polished forms. Good charts are rarely produced in a single step; they emerge through iteration, critique, and reflection.

Finally, we considered how contemporary tools—from spreadsheets to AI-assisted visualization—have transformed both the speed and the stakes of visual communication. While these technologies lower barriers to entry and accelerate insight, they also heighten ethical responsibility. Automation does not replace judgment. Scientific communicators must remain accountable for every design choice, verify outputs, disclose assistance, and prioritize trust over convenience.

Taken together, this chapter argues that effectivedata visualization is not about mastering software or following rigid rules. It is aboutmaking informed, ethical choices that respectboth the data and the audience. When designers ground their work in clarity, honesty, andempathy—and treat technology as a collaborator rather than an authority—visualizations can do what they have always done at their best: make complex realities visible, intelligible, and meaningful.

APPENDIX

COMMON CHART TYPES IN SCIENCE COMMUNICATION GLOSSARY

Chart Type — **Best Use and Strategy**

Bar Chart

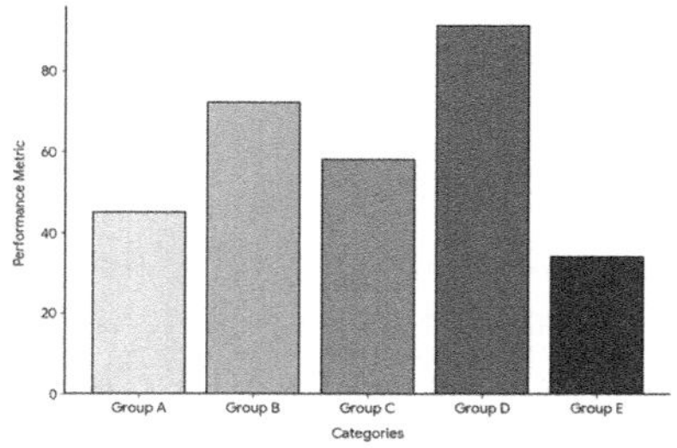

Comparing Groups: Ranks discrete categories. **Tip:** Use horizontal bars if labels are long to avoid diagonal text.

Line Chart

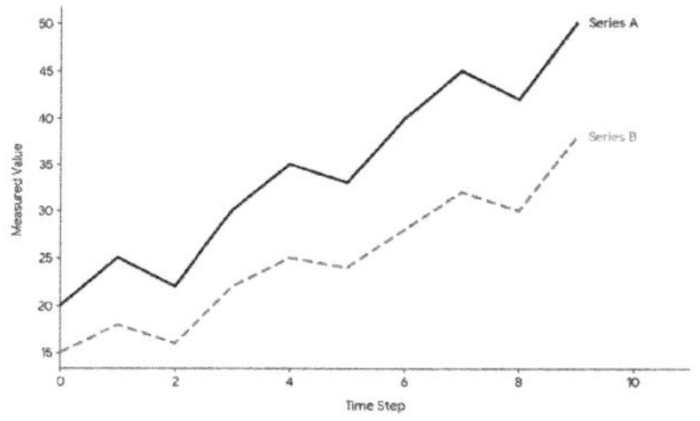

Temporal Trends: Tracks changes over time. **Tip:** Remove legends and use direct labels at the end of lines to reduce eye travel.

Pie Chart

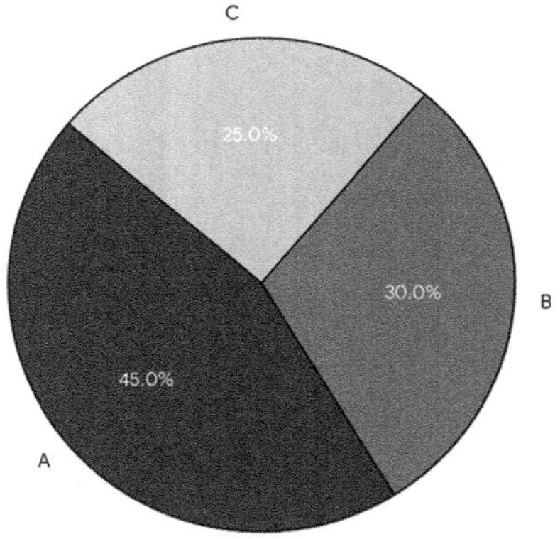

Parts-of-a-Whole: Shows simple proportions. **Tip:** Limit to 3–4 slices; for complex data, a **Stacked Bar** is more precise.

Scatterplot

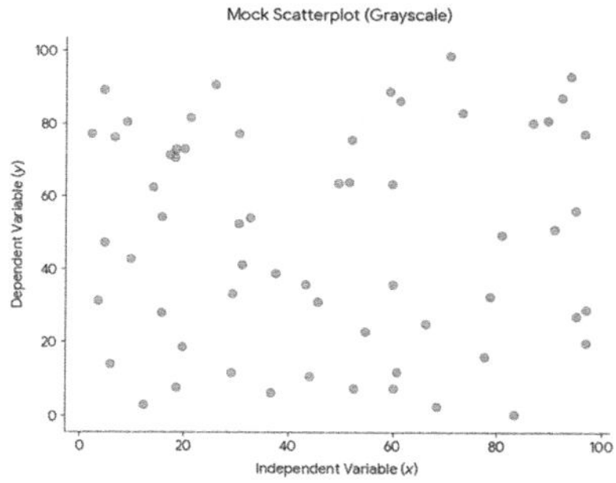

Correlation: Identifies relationships between two variables. **Tip:** Add a **Trend Line** (regression) to help viewers interpret the link.

Chart Type		Best Use and Strategy
Bubble Plot	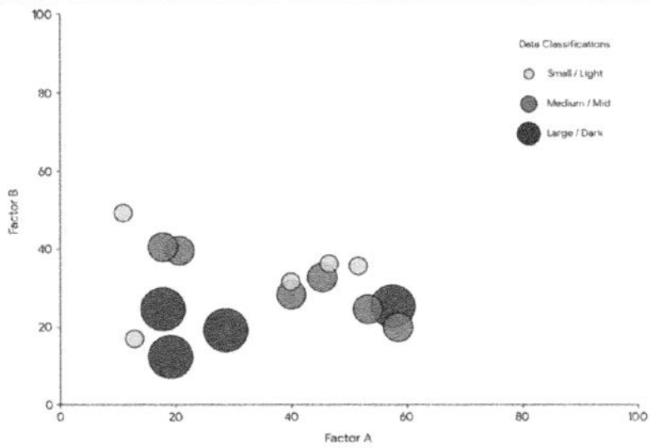	**3-Variable Relationship:** Adds "size" to a scatter plot. **Tip:** Ensure the bubble **area** (not diameter) represents the value.
Histogram	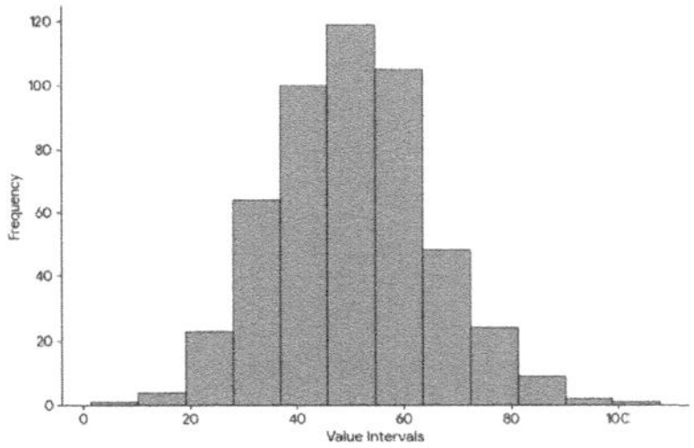	**Data Distribution:** Shows the frequency of values in "bins." **Tip:** Adjust bin size; too many create noise; too few hide the trend.
Box Plot	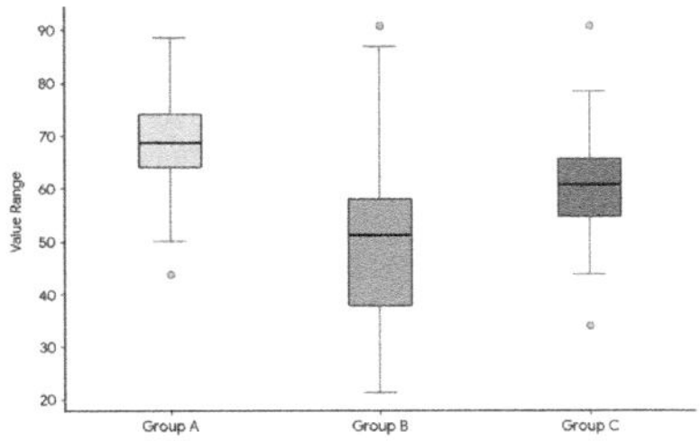	**Statistical Spread:** Summarizes range, median, and outliers. **Tip:** Include a small "How to Read" key for non-technical audiences.
Area Chart	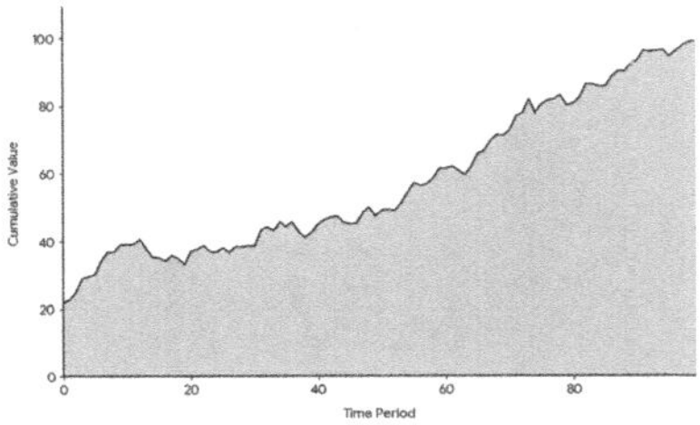	**Cumulative Volume:** Emphasizes the total "mass" under a trend. **Tip:** Use transparency (alpha) when layering multiple series.
Heat Map	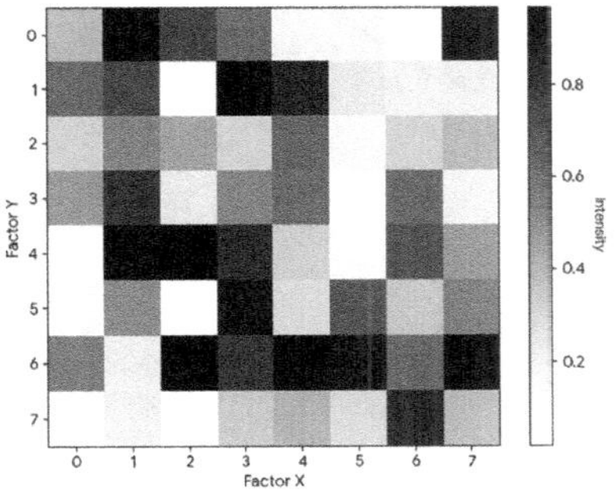	**Density Patterns:** Visualizes intensity within a grid. **Tip:** Use a colorblind-safe palette or grayscale gradients.

<table>
<tr><th>Chart Type</th><th></th><th>Best Use and Strategy</th></tr>
<tr><td>Slope Chart</td><td>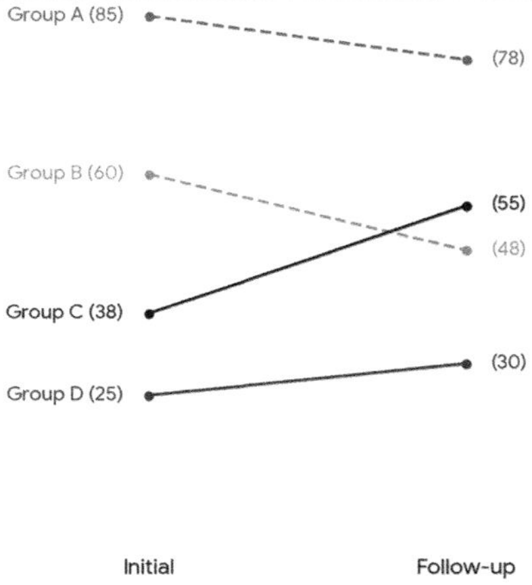
</td><td>Comparative Change: Highlighting “Before vs. After” shifts. Tip: Use black for increases and gray for decreases to show impact instantly.</td></tr>
<tr><td>Sankey</td><td>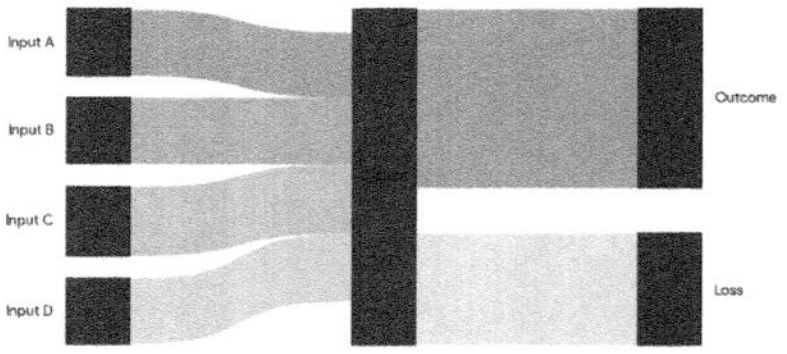
</td><td>Flow & Attrition: Tracks volume movement through a system. Tip: Keep flows moving in one direction (Left ⍰ Right) to ensure clarity.</td></tr>
<tr><td>Chord Diagram</td><td>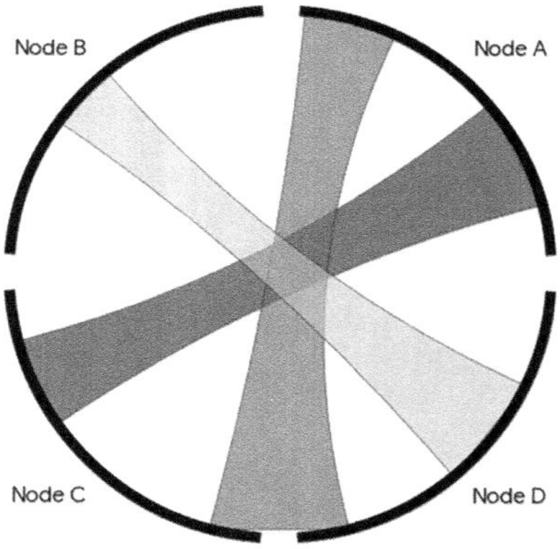
</td><td>Complex Interconnectivity: Shows reciprocal flows or system logic. Tip: Use minimalist icons to focus on the “how” of the system.</td></tr>
</table>

Chart Type		Best Use and Strategy
Conceptual Diagram	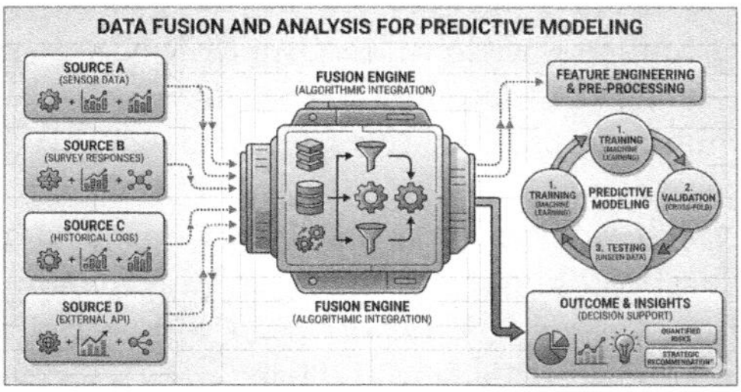	**Systems and Logic:** Visualizes theoretical frameworks, abstract relationships, or non-numerical processes. **Tip:** Use minimalist icons and clear directional arrows to guide viewer through the narrative flow.

REFERENCES

American Museum of Natural History. (2023). *Human population through time* [Video]. New York, USA. https://www.amnh.org/explore/videos/humans/human-population-timeline

Beninger, J. R., & Robyn, D. L. (1978). Quantitative graphics in statistics: A brief history. *The American Statistician*, *32*(1), 1–11.

Berinato, S. (2016). *Good charts: The HBR guide to making smarter, more persuasive data visualizations*. Harvard Business Review Press.

Bertin, J. (1967). *Sémiologie* graphique: Les diagrammes, les réseaux, les cartes. Gauthier-Villars.

Cairo, A. (2013). *The functional art: An introduction to information graphics and visualization*. New Riders.

Cairo, A. (2019). *How charts lie: Getting smarter about visual information*. W. W. Norton & Company.

Descartes, R. (1637). *La géométrie*. Chez Ian Maire.

Dilke, O. A. W. (1987). The culmination of Greek cartography in Ptolemy. In J. B. Harley & D. Woodward (Eds.), *The history of cartography: Vol. 1. Cartography in prehistoric, ancient, and medieval Europe and the Mediterranean* (pp. 189–210). University of Chicago Press. https://press.uchicago.edu/books/hoc/HOC_V1/HOC_VOLUME1_chapter11.pdf

Few, S. (2011, June). The chartjunk debate: A close examination of recent findings. In *Visual Business Intelligence Newsletter*. https://www.perceptualedge.com/articles/visual_business_intelligence/the_chartjunk_debate.pdf

Knaflic, C. N. (2015). *Storytelling with data: A data visualization guide for business professionals*. Wiley.

Lamott, A. (1994). *Bird by bird: Some instructions on writing and life*. Anchor Books.

Minard, C. J. (1869). *Figurative map of the successive losses in men of the French Army during the Russian campaign 1812–1813* [Map]. https://commons.wikimedia.org/wiki/File:Minard.png

Nightingale, F. (1858). *Diagram of the causes of mortality in the Army in the East* [Diagram]. https://commons.wikimedia.org/wiki/File:Nightingale-mortality.jpg

Oresme, N. (1505). Tractatus *de latitudinibus formarum* [Digital image]. Wikimedia Commons. https://commons.wikimedia.org/wiki/File:Oresmes_diagrams_full_page.gif

Playfair, W. (1786). *Exports and imports to and from the East Indies from the year 1700 to 1780* [Chart]. Wikimedia Commons. https://commons.wikimedia.org/wiki/File:1786_Playfair_-_3_Export_%26_Import_to_and_from_the_East_Indies_from_the_year_1700_to_1780.jpg

Roam, D. (2008). The back of the napkin: Solving problems and selling ideas with pictures. Portfolio.

Rosling, H. (2006). *The best stats you've ever seen* [Video]. TED Conferences. https://www.ted.com/talks/hans_rosling_the_best_stats_you_ve_ever_seen

Snow, J. (1854). *Map showing the distribution of the cases of cholera in the district of Soho* [Map]. Wikimedia Commons. https://commons.wikimedia.org/wiki/File:Snow-cholera-map-1.jpg

Snow, J. (1854). *On the mode of communication of cholera*. John Churchill. https://collections.nlm.nih.gov/catalog/nlm:nlmuid-0050707-bk

Tufte, E. R. (2001). *The visual display of quantitative information* (2nd ed.). Graphics Press.

Tukey, J. W. (1977). *Exploratory data analysis*. Addison-Wesley.

Index

Note: **Bold** page numbers refer to tables and *italic* page numbers refer to figures

For Product Safety Concerns and Information please contact our EU representative GPSR@taylorandfrancis.com Taylor & Francis Verlag GmbH, Kaufingerstraße 24, 80331 München, Germany

Batch number: 10783758

Printed by Printforce, the Netherlands

Communicating Science in Public

Amid intersecting climate and health crises, as well as rapid developments in artificial intelligence and new media forms, the need for communicating science with nuance and care has never been greater. *Communicating Science in Public* supports science communicators who seek to engage in compelling ways with audiences with varied levels of scientific expertise. The volume provides not a rule book, nor a set of templates and formulas, but rather a portable set of rhetorical strategies for working across science communication genres with the expectation that these genres will continue to change and new ones will emerge.

Introductory chapters provide a foundation for communicating science in public across genres. The genre-specific chapters—including chapters on personal essays, op-eds, science exhibits, and podcasts—then offer histories of how a given genre evolved as well as where it fits into broader media ecologies. These chapters provide examples of the genre, key considerations for navigating its affordances and constraints, and a process for brainstorming, composing, revising, and reflecting.

By taking a genre-based approach, this book asks its readers to reimagine science communication as more than an act of translating complex ideas into simpler terms. Instead, readers learn to reflect on the opportunities and limitations of different forms of communication and how to engage diverse publics in inclusive, accessible ways, developing transferable, transformative rhetorical knowledge.

This book can be used as a science communication textbook geared toward advanced undergraduate students and graduate students with a foundation in the sciences who are learning how to communicate scientific ideas to non-specialist audiences.

Key Features:

- Helps science communicators navigate a complex communication landscape.
- Builds skills for engaging public audiences through multiple genres and modalities beyond print.
- Offers practical advice for scientists, new and established, who want to communicate their science more broadly.